Information Communication Technology and Social Transformation

This book argues that information communication technologies are not creating new forms of social structure, but rather altering long-standing institutions and amplifying existing trends of social change that have their origins in ancient times. Using a comparative historical perspective, it analyzes the applications of information communication technologies in relation to changes in norms and values, education institutions, the socialization of children, new forms of deviant and criminal behaviors, enhanced participation in religious activities, patterns of knowledge creation and use, the expansion of consumerism, and changing experiences of distance and time.

Hugh F. Cline received his Ph.D. in Sociology from Harvard University in 1967. For 20 years, he served as Executive Research Director at Educational Testing Service. Currently, he is an Adjunct Professor at Teachers College, Columbia University. His research has focused on the use of information communication technologies in complex organizations.

Routledge Studies in Science, Technology and Society

Information Communication Technology and Social Transformation

A Social and Historical Perspective

Hugh F. Cline

Routledge
Taylor & Francis Group

NEW YORK   LONDON

First published 2014
by Routledge
711 Third Avenue, New York, NY 10017

and by Routledge
2 Park Square, Milton Park, Abingdon, Oxon OX14 4RN

*Routledge is an imprint of the Taylor & Francis Group,
an informa business*

© 2014 Taylor & Francis

The right of Hugh F. Cline to be identified as author of this work has been asserted in accordance with sections 77 and 78 of the Copyright, Designs and Patents Act 1988.

Library of Congress Cataloging-in-Publication Data
Cline, Hugh F.
   Information communication technology and social transformation : a social and historical perspective / by Hugh F. Cline.
      pages cm. — (Routledge studies in science, technology and society ; 25)
   Includes bibliographical references and index.
   1. Information technology—Social aspects.   I. Title.
      HM851.C586 2014
      303.48'33—dc23
      2013043925

ISBN13: 978-1-138-01680-4 (hbk)
ISBN13: 978-1-315-78067-2 (ebk)

Typeset in Sabon
by IBT Global.

To Hilary for her continuing support and love.

Contents

Figures

Preface

This book uses historical and sociological perspectives to examine social change in the context of the uses of contemporary information communication technologies (ICTs). It is intended to engage readers who are interested in a broad survey of how computers and networks are bringing about change in society. The book provides a historical context in which to consider the many ways that these technologies are influencing ongoing patterns of social change. In addition, this book will be useful as supplementary reading in undergraduate and post-baccalaureate courses in the social and behavioral sciences.

A website for this book will be established at the time of publication. As so many of the examples in the text refer to recent development and innovations, it will be prudent to maintain a website that will be updated on a monthly schedule to alert readers to news pertinent to the major themes of social change as influenced by ICTs.

The format and style in which this book is presented are not those usually followed in a scholarly book. Rather than using footnotes and citations of materials that are found in the mass media and the scientific literature, each chapter in the book has appended brief bibliographical notes that will identify the sources of the major materials covered in that chapter. In addition, mention of books, articles, and websites are included for those readers who might wish to pursue further some of the points discussed in the text.

Acknowledgments

This book has been prepared over a sixteen year period, and it is impossible to acknowledge my indebtedness to all the many people who have contributed to its completion. First, I would like to acknowledge the immense contributions of all the Teachers College students who since 1999 have participated in my course, Technology and Society. They have all been exposed to the intellectual content of this volume, and many have offered valuable suggestions for strengthening various drafts. In my judgment this interaction represents the most valuable aspects of the teaching and learning activities of higher education, and I thank all the students for their assistance.

A number of Teachers College colleagues have been of enormous assistance in facilitating the intellectual growth and productivity of this Adjunct Professor of Sociology and Education. Foremost, I wish to acknowledge the invaluable assistance and support of Gary Natriello whose advice and encouragement are most appreciated. In addition, the support of colleagues Robert McClintock, the late Frank Moretti, Charles Kinzer, Allen Foresta, and Diane Katanik has been invaluable.

In any intellectual effort that spans so many years the support of family members is critical to insure completion. For their forbearance and good humor during this time I wish to thank my son, Hugh Cline, and step-children Beth Ogilvie Freda, Mark Freda, Bob Ogilvie, Bill Ogilvie, and Brad Ogilvie. All have been most gracious. Particularly, I wish to acknowledge and thank my daughter, Lynn Cline, an author and journalist, for her skill in transforming the ponderous writing style of an academic trained sociologist into a much more acceptable format. The application of her fine Italian hand has vastly improved this volume, and I am eternally grateful.

And lastly, I acknowledge the patience and loving support of my wife, Hilary Hays to whom this book is dedicated.

1 Introduction

SCOPE OF INQUIRY

Computers and networks have generated an enormous amount of controversy and debate. Depending on whom you ask, computers and networks either:

> create political turmoil, or promote democracy;
> invade privacy, or promote transparency;
> increase social isolation, or expand social and professional networks;
> provide new opportunities for minorities, or create a digital divide;
> destroy literacy, or produce new forms of creative expression;
> make us stupid, or increase our cognitive capacities;
> encourage unnecessary purchases, or expand the consumer economy;
> enable greater exposure to media, or encourage copyright violations;
> assist law enforcement agencies, or enable new forms of crime; and
> eliminate teachers' jobs, or improve teaching and learning.

Many journalists, pundits, and social scientists have recently made all these somewhat contradictory statements. Some evidence supports each one, but very few are based on valid and reliable research. Most are the result of limited observations. Such statements contribute to a great deal of confusion and add to the general public's misunderstanding of the many changes that are occurring in our society. Indeed, social change is escalating at an increasing rate, and computers and networks are major contributors to those changes, particularly in the past 50 years. Yet most of the public discourse lacks adequate consideration of both historical and sociological perspectives. This book is an attempt to provide those perspectives.

Following the contemporary practice of using a lowercase "e" as a prefix, *eSocial Change* might have been used as the title of this book. This format is increasingly used to denote that the following word or phrase refers to something taking place in an *electronic* environment. For example, email, ebook, and elearning convey that mail is transmitted over the Internet, a book is read on a computer screen, and learning is conducted via an interactive network. In addition, many terms use the prefix "i." For example, iPhone, iTunes, and iPad all refer to systems that enable *interactions* via connections to networks.

Increasingly, much of our lives are taking place in "e" and "i" environments mediated by systems enabled by computer capabilities. Most people are aware that new electronic capabilities are becoming commonplace in our lives, but they do not recognize that many of these changes are not fundamentally new. Rather, they are continuations of long-term trends of social changes that date back to the beginning of recorded history. The objective of this book is to demonstrate the continuity of those lines of historical social change and examine them in the context of the information communication technologies (ICTs) that increasingly permeate our world, affect our lives, and influence how we see ourselves as well as others.

It has been almost four decades since the late Daniel Bell, the eminent sociologist, analyzed trends of social change and described the emerging post-industrial society. According to Bell, this new social order would be founded upon the then-emerging information technologies. He foresaw a future in which production would gradually be automated by machines, and a new economy would emerge in which information drove the provision of services. Furthermore, economic and political power would be based on the knowledge and the expertise of the information industries. Henry Ford and the automobile industry were the icons of the industrial era, and Bill Gates and the Internet personify the post-industrial one. In the intervening years since the publication of Bell's prescient book, many of the trends he predicted have come to fruition. The time has now arrived to examine systematically how the distribution and utilization of information technologies are influencing the major trends of social change. This book is a first attempt to examine patterns of historical social change in the context of the growing use of ICTs for the creation, processing, storage, retrieval, dissemination, and use of knowledge. A broad definition of ICT is employed in this book, and three terms, *information, communication,* and *technology,* describe the focus of these explorations. Metaphorically speaking, information is the content, communication is the process, technology is the vehicle, and social change is the outcome.

Since the Industrial Revolution, society has been undergoing significant transitions at an escalating rate. All the major social institutions, including the family, the economy, religion, polity, and the law, are constantly changing, adapting, and evolving into new forms. Within just the past half century, ICTs have become pervasive in society, and they are exercising profound influences on major social transformations. This book examines both the obvious and more subtle patterns of change brought about by ICTs. In addition, the book speculates on the future direction of some of these transitions. This inquiry seeks both comprehensiveness and comprehension in the examination of social change. Comparisons across the major institutions of society will reveal new insights pertaining to social change.

DEFINITIONS

Initially, it is important to be clear about the boundaries of these inquiries and declare what is and is not included in the concept of ICTs for the purpose of analyses in this book. First, the definition of each of the three words in the phrase *information communication technology* is explored, and then they will be combined to focus the inquiry of this book. According to the *Oxford English Dictionary*, the word *information* is derived from the Latin word *informare*, meaning "to inform, instruct, or teach." By the late Middle Ages, the word had come to mean "knowledge or facts communicated about a particular subject, event, etc." The *OED* now defines information as: "Without necessary relation to a recipient: that which inheres in or is represented by a particular arrangement, sequence, or set, that may be stored in, transferred by, or responded to by inanimate things." There are three important components to this definition.

First, information has no necessary relation to any recipient. This implies an expectation that there may be multiple contributors to, users of, and facilitators of the use of information. Global access and use are hallmarks of many current ICT systems. Second, the inherent meaning of information is represented in a symbolic form by arrangement, sequence, or set to convey some fact or knowledge. The conveyance of fact or knowledge implies the older definitions of informing, instructing, or teaching. Symbolic representation of numbers, letters, and characters is one of the fundamental design features of ICT systems. And third, inanimate objects may be used for storing, transferring, and responding to information. These operations are the basic functions of electronic computers. Whether computers are inanimate can be debated. Certainly the boxes that house computers do not move. It is true that electrical pulses move within computers of all sizes. However, when we consider robots driven by computers, inanimate certainly does not pertain. Nevertheless, the close relationship between the definitions of information and computer technology is obvious.

The *OED* lists 11 definitions of the word *communication*. Older meanings are related to the word *common*, and implications of sharing appear in some of the earliest recorded uses of the term. However, among contemporary definitions, the one that is most appropriate for this discussion defines *communication* as the "imparting, conveying, or exchange of ideas, knowledge, information, etc., whether by speech, writing, or signs. Hence, [communication is] the science or process of conveying information, esp. by means of electronic or mechanical techniques." Information and communication are the hallmarks of literate societies and have existed for thousands of years.

The history of the definition of the word *technology* is both more complex and interesting. The word is derived from the Greek *techne*, which originally meant "art," as in the sense of a craft. During the Middle Ages, the concept evolved to include the craft skills possessed by artisans, known as applied art. These skills for working with tools or simple machines were learned in apprenticeships. Subsequently, the concept began to imply possessing knowledge or expertise in applied arts and sciences. By the late eighteenth century, the word *technology* itself emerged, referring to the study of the science or discipline of techniques. The *OED* defines technology today as "a particular mechanical art or applied science."

The *OED* includes a definition for the phrase *information technology* as "The branch of technology concerned with the dissemination, processing, and storage of information, esp. by means of computers." The phrase *information technology* is widely used to refer to computers of all sizes, whether they operate as stand-alone machines or are tied into large networks. These machines initially were used primarily to process and analyze numerical data. An important impetus in the U.S. to the development of computers during World War II was the calculation of gunnery or firing tables for military cannon. That war had ended by the time computers were sufficiently sophisticated to calculate the tables, but for many years thereafter computers were used almost exclusively for numerical data processing and analysis. Many astute people realized that computers were, in fact, general purpose symbol-processing machines that could analyze patterns of both numbers and letters, thereby enhancing a wide variety of mathematical and logical problem-solving operations. During the decades following the war, many computer programs were developed to perform large and complex data-processing and analysis tasks on both numerical and textual data. However, it was not until the 1980s and 1990s that computers were used to form extensive networks, such as the Internet. At that time, communication began to emerge as a dominant feature of the technology. Hence, the phrase *information communication technology* has now come into common use.

ICT refers to at least two kinds of activities. First, it is frequently used by international organizations such as the United Nations, the World Bank, or the Organization for Economic Cooperation and Development to describe the efforts they promote and support among developing nations to participate more fully in the global economy. ICT also refers to the hardware, software, and infrastructure that support the communications necessary for a nation to function as an effective economic participant in international markets.

A second common use of the phrase ICT refers to the skills that individuals need to function in an information-rich society. The Educational Testing Service (ETS) has recently developed, field tested, and is now selling a new instrument to assess an individual's proficiency in ICT skills. ETS defines this proficiency as "the ability to use technology as a tool to research, organize, evaluate and communicate information, and the possession of a fundamental understanding of the ethical/legal issues surrounding the access and use of information." This test, named iSkills, is designed to assess students' proficiency as they enter college. The iSkills test should not be confused with ETS's long-standing college admissions examination, the

Scholastic Achievement Test (SAT). ETS anticipates that iSkills test results will be used for individual remediation, program evaluation and reform, and international comparisons. Clearly, the domain included in the new ETS instrument is broader than knowledge of computers, for it encompasses the ethical dimensions of the analysis, utilization, and communication of information.

This broader definition of ICT is employed in this book. Together, *information*, *communication*, and *technology* encompass the foci of this analysis. An additional note will clarify further the scope of this inquiry. An examination of evidence will address whether the structure and function of major societal institutions are changing in fundamental ways as a consequence of widespread employment of ICTs for the creation, processing, storage, retrieval, dissemination, and utilization of information.

The acronym ICT is frequently used by the mass media as well as in some scholarly literature as the functional equivalent of another commonly used phrase: the *information revolution*. Although it is clear that the production of new information is escalating rapidly, calling this a revolution is an overstatement. The storage, retrieval, transmission, and processing of all this new information is the foundation of the now substantial ICT industry, an industry whose growth was initiated and is now sustained by computers. It needs to be stressed that information technology as defined here hardly existed only 60 years ago. It helps to understand this so-called revolution by considering it in a broader historical context. The history of the technology of information transmission as a cultural change agent extends back more than 3,000 years, from clay tablets, papyrus scrolls, pen and ink, printing presses, telegraphs, radio, film, and television up to contemporary mobile computers and networks. From the perspective of the ever-changing technology, it is premature to employ the term *revolution*. However, the phenomenal growth in the number of people who use computers and networks for communication certainly qualifies as a major social change.

As an additional preparatory note, this book employs a systemic perspective. The term *systems analysis* has been used to describe a wide variety of methods for examining change. Although the term may bring to mind some complex mathematical model that simulates the interactions of economic processes, systems analysis is really quite an ancient and simple concept. A *system* is any collection of phenomena that are related or connected to form a more complex entity, and *systems analysis* is the examination of the interactions among those components over some period of time. Another definition involves the investigation of interrelatedness. Societies are systems, for the various institutions that comprise a society (i.e., family, education, and polity) are interrelated. This book examines how ICTs are being implemented in major societal institutions. It also demonstrates how these institutions are increasingly interrelated. Therefore, cross-references to analyses in different institutions are common in the following discussions.

FOCI

In this introductory chapter, it is useful to point out what this book is not. First, it is not an attempt to predict how society will function in the future as a result of the increasingly widespread dissemination and utilization of ICTs. Prognosticators of all persuasions, including scholars, futurists, technophiles, and pundits alike, are all enamored with their own attempts to see into the future and engage our imaginations with vivid scenarios of what society will be like in coming decades. The reading of such books can be amusing. They are rarely based on careful analyses, and they quickly become obsolete, out of print, and remaindered.

What is offered here differs greatly from predicting the future. Rather, this book identifies patterns and trends in how the uses of ICTs are affecting changes in our major societal institutions. There are no initial assumptions that social institutions are being altered with either positive or negative consequences. The daily press is replete with stories of new developments in how computers and networks are distributed and used. However, this exploration begins with a healthy degree of skepticism as to whether these changes may be beneficial or detrimental. With a systematic exploration, a global assessment of the extent of societal changes is possible. Furthermore, the effort illustrates some of the underlying dynamics with which ICTs may possibly affect social changes that have been in process for many years. Rather than striving for the unattainable goal of currency, this exploration seeks both comprehensiveness and comprehension within a historical and sociological perspective. By making historical comparisons across the major institutions of society, new insights may emerge.

Almost all the accounts of ICT applications reviewed in this book have been reported in recent years in the popular media and scholarly journals. What will be original and hopefully valuable here are the insights gained from looking across the major institutions in society for evidence of patterns and trends in changes in structure and function that are associated with ICT applications. Special attention is paid to topics such as child-rearing patterns, participation in political activities, government service delivery, leisure time activities, and forms of religious expression.

This book is not about all forms of technology. Information is the sole focus of this book. Biomedical technology and genetic engineering are other forms of technology that become increasingly pervasive. The consequences of their applications are already widespread and profound. The nature of these technologies and the kinds of applications are fundamentally so different from ICTs that they will not be included in these analyses, because the scope of this inquiry would become unmanageable. If anything, criticism might be directed at this effort for being too inclusive. However, this is the first attempt to examine the major social institutions to identify common patterns of accommodation to the emergence of ICTs.

One further comment pertaining to scope is necessary. In all the discussions that follow, the focus is primarily on what until recently was called Western Civilization, which used to include Europe and North America. A more contemporary way to describe the scope is to say that it includes nations that are transitioning from an industrial to a post-industrial or informational state, including Japan, South Korea, and Finland, three nations where ICTs are widely distributed and used. All these geopolitical areas provide the most fruitful sites to observe relationships among social structures and ICT applications. It is worth observing that the proliferation of ICTs in the so-called developing nations is also rapidly escalating. The growth rates of the use of cell phones and the Internet in China, India, and African nations are phenomenal; and future cross-cultural comparative analyses promise to be extremely productive.

A cursory historical review of the development of ICTs, ranging from large mainframes to minicomputers, personal computers, laptops, networks, and most recently smart phones, covers a span of approximately six decades. Projections of where technology will take ICTs in the future include peta scale computing, in which ten raised to the 15th power operations can be performed in one second; organic computing, in which biological materials are used for input, output, and computation; and quantum computing, a technology based on principles of quantum theory. Rather than engage in that kind of guesswork about the future of computing, the following chapters will examine the underlying changes in social institutions that reveal modifications of long-standing patterns of societal change. The task of guessing what technology will next be developed is left to others. What is of interest in this book is discerning social changes that have occurred in the past, particularly in the last half century, that may be harbingers of trends of future societal forms.

BOOK PREVIEW

It is always prudent to provide readers with chapter previews or an itinerary of a coming intellectual journey. Chapter 2 presents a review of three sociological perspectives of societal evolution that frame the explorations of social change and ICT applications. The first category focuses primarily on the location and sizes of human societies, progressing from nomadic tribes to agricultural communities to large metropolitan areas and eventually to a globalized planet. The second category focuses on the evolution from preliterate to literate societies, then to information societies, and eventually to a mass and multi-media society in which iconography begins to emerge as a mode of communication. The third category encompasses theories that address the economic and political dimensions of societal evolution from the initial focus of providing food and shelter to mass production and consumption and then to an economy structured on knowledge and expertise.

This third category includes a continuing focus on maintaining social order both within societies and across modern nation-states. In discussing these three categories, brief references are made to some of the writings of selected social theorists who have examined societal evolution.

Patterns of change in the normative order are the foci in Chapters 3 and 4. Normative order consists of the shared values and expected behaviors that account for the vast uniformities of the activities of individuals and groups in society. The shared values are incorporated in the norms, mores, folkways, common practices, rituals, and customs that guide human behavior. Many aspects of the normative order are formally stipulated in rules, standard operating procedures, administrative laws, judicial rulings, and legislation. In the absence of normative order, social chaos ensues. All societies strive to maintain order and minimize chaos. It should be pointed out that normative order is not a static phenomenon. Values, norms, and laws are constantly changing. Some aspects of normative order change slowly; others change quickly. Normative order, then, is a constantly evolving set of asynchronously shifting values. When rapid changes occur in society, the normative order almost always lags behind. Conflicts frequently arise between those who champion the new norms and those who resist the changes. The current debates over globalization reflect some of these patterns. Chapter 3 examines changes prompted by ICT applications in the patterns of social inequality in the division of labor. It also examines the role ICTs play in changing patterns of government and political activities. Chapter 4 takes up the issues of intellectual and property ownership, individual and organizational privacy, the role and practice of religion, and new forms of deviance and crime.

The changing patterns of socialization and education are the topics analyzed in Chapter 5. At birth, all humans are capable of a vast array of behavioral acts. Yet, over the course of a person's life, only a small subset of those potential acts is ever executed. In the process of growing from an infant to a fully independent adult, each person becomes aware of and, in most instances, abides by the prevailing normative order. Through mimicking or modeling the behavior of significant adults, children learn or become socialized to what most other people recognize as appropriate and inappropriate behaviors. From the beginning of the Industrial Revolution, patterns and agents of socialization have undergone substantial modification. In agricultural societies, extended families were the principal agents of socialization for children. Parents, older siblings, and grandparents guided children to develop their value systems and their accompanying behavioral repertoire. In many societies, the church also played a major role in socialization.

As the Industrial Revolution progressed and schools opened to provide an educated labor force, teachers became important socialization agents. At the same time, nuclear families became more common, and the presence and influence of members of the extended family diminished. More recently, female-headed households consisting of mothers and sometimes grandmothers, unmarried couples, same-sex pairs, and various other forms of blended families have taken on the responsibility for childhood socialization. Concern is growing that the role of the primary agents in socialization is shifting away from the family, and the values of the popular culture are replacing those of the dominant normative order. ICTs play a prominent role in the dissemination of that popular culture. Furthermore, childhood and adolescent peers increasingly use social software enabled by ICTs to reinforce and expand that popular culture. Note that these concerns are based primarily on anecdotal evidence; as yet, little systematic social science research has examined the influence of ICTs in the context of families and socialization.

Chapter 5 also takes up the changing patterns of education. One major change in modern societies is that more people of all ages participate in education and training to be effective participants in both the labor force and other adult roles. Indeed, many people now recognize the need to prepare children to be lifelong learners. This chapter explores the relationship between ICTs and changing patterns in education from four perspectives: changes in pedagogy or methods of teaching and learning, changes in the curriculum to produce students who will be effective users of ICTs throughout their lives, the extraordinary growth of online educational programs and institutions, and financial problems of schools and institutions of higher education.

The explosion in the creation, dissemination, and utilization of knowledge is a major trend common to post-industrialized societies. ICTs enable this expansion and are influencing the development of new modes of creating and using knowledge. Indeed, many people characterize the post-industrial society as being founded on knowledge, rather than products or services. Chapter 6 describes examples of such changes in the knowledge industries, including research and development organizations, colleges and universities, research and public libraries, publishing houses, newspapers, magazines, and broadcast and cable television companies. The chapter then explores these contemporary developments in the context of broader historical patterns that are rooted in the various transformations of literacy.

Chapter 7 explores the trend of increasing consumerism in modern societies. Ever since the Industrial Revolution, the economies of most nations have struggled to sustain ever-growing markets. A commonly accepted notion is that expanding markets portend a healthy economic future. Mass production, marketing, and consumption are the hallmarks of developed nations. In each of these three domains, ICTs play an increasingly pivotal role. The Internet has become a major driving force in marketing and consumption. This chapter discusses how ICTs are employed to develop new methods of increasing economic growth, particularly through online consumption. Web advertising is rapidly growing and is diminishing traditional print and television revenues. Patterns of consumption are likewise changing with online shopping. The concept of *niche*, or finely targeted, advertising, marketing, and purchasing is replacing the older model of consumerism as embodied in the full-service department store. ICTs are enabling these new strategies.

Although the topics presented so far have been discussed by social scientists, the objects of inquiry in Chapter 8, social time and social space, have received less attention. As more social scientists begin to recognize the growing influence of ICTs, the social use of time and space will come under greater scrutiny. The perception and use of space are changing as a consequence of increasing proliferation of ICTs. In many settings, physical space is being complemented by virtual space or cyberspace. Some scholars are even discussing the death of physical space. Although that seems extreme, it is now easy to engage in interactions or conduct business around the globe without regard to distance or local time. This capability has profound influences on where and when people work and how they allocate their time among job, family, and leisure. Indeed, there is some evidence that leisure time is disappearing in some locales, and people are having difficulty getting adequate actual and virtual distance from work. This chapter explores three topics: altered perceptions, a brief history of social time and distance, and virtual communities.

The final chapter, Chapter 9, summarizes the major findings and conclusions of the book's analyses. It discusses topics related to future social changes that are likely to be influenced by the further proliferation of ICTs: the substance and methodology of policy-oriented research using large databases, new forms of literacy, the appropriate definition and implementation of transparency in both the public and private sectors, culture wars, and the extent to which globalization and the new mass media can promote world peace and harmony across many different cultural and ethnic groups.

Lastly, two appendices are included. The first is a brief introduction to the design and architecture of digital computers, and the second is a short history of the development of computer networks. Both are presented for readers who are unfamiliar with these topics to assist in understanding why, for example, online maintenance of bank accounts is a relatively simple task for networked computers to accomplish. The activities of a large-scale, resource-sharing system such as a bank can be broken down into the steps of basic computer instructions. Each and every step in the process of keeping track of the flow of monies or communications is known and can readily be coded as instructions in a computer program. However, creating a computer program to function as a teacher is enormously more difficult, if at all possible. The process of teaching calculus to a student requires many global and subjective assessments of the learner's prior knowledge and understanding, relevant pedagogical strategies, and appropriate feedback on performance. No one knows yet how to specify such instruction and assessment procedures in an algorithm that is amenable to the creation of a computer program. The appendices point out that the basic instructions of computers execute only the simplest arithmetic and logical operations. Understanding this distinction between a banking system and a calculus teacher is critical for understanding some of the major themes of this book.

2 Societal Evolution

The history of the evolution of society has always interested scholars. How society might evolve in the future is of equal interest. Countless scholars have suggested different perspectives for understanding this evolution and the driving forces in the history of civilization. Most of these theories refer exclusively to Western Europe and North America, and a broader historical and cultural perspective is needed. Nevertheless, these theories address various aspects of societal evolution over the recorded history of Western Civilization and will be useful in the explorations of information and communication technologies (ICTs) and their effect on the trends of change in major societal institutions.

This brief chapter provides a background for the examination of social change in the context of the ever-expanding use of ICTs. For those readers who are already familiar with the social science literature on change, this chapter serves as a review and refresher experience. For the majority of readers, it is a brief summary of the major theoretical schools of social change. It also provides the background for subsequent chapters.

LOCUS

The first category of societal evolution theories examines the locus or physical site of human communities. The archeological and historical records clearly indicate a major transition in the spatial settings of societies—humans evolved from nomadic hunter and gatherer societies to ones in which sustained agricultural activities occurred in a fixed location. Perhaps the farming community might move to another site when fertile soil was depleted or climate changes occurred, but relatively stationary agrarian communities gradually replaced the nomadic pattern. As agricultural activities became more efficient and food supplies could be accumulated, not everyone needed to participate in farming; some people were free to pursue other activities, such as tool making or food processing. This type of specialization of activities and then the exchange or barter for goods or services formed the foundation of the division of labor in society.

Gradually, as small communities were established, the division of labor became more complex. Needs emerged for new activities, and occupations such as millers and blacksmiths were created to meet those needs. As industrialization progressed and urban areas began to appear, organizations such as factories, banks, shipping companies, and government regulatory agencies were created to organize and monitor those functions. The division of labor began to take on some of the characteristics recognized today in large-scale bureaucracies. Such communities became the foundation from which modern urban life emerged. Many social theorists describe the transitions from nomadic to rural and agrarian to urban as the defining characteristics of societal evolution.

The work of two sociologists, Ferdinand Toennies (1855–1936) and Emile Durkheim (1858–1917), further illuminate these transitions. Both wrote in the early twentieth century. Toennies's famous distinction between *gemeinschaft* and *gesellschaft* highlights the transition in societal organization from the smaller, agrarian communities to the modern, industrial, urban society. *Gemeinschaft,* or "community," as it is usually translated from the German, is characterized by family and neighborhood ties. Its members share a common set of values and norms, and its essence is found in a rural, agricultural setting. *Gesellschaft*, or "society," on the other hand, is characterized by formal and usually financial ties. Its members share only the values and norms that enable them to achieve their common goals and objectives, and its essence is found in urban settings.

Durkheim makes a similar distinction, but he uses the terms *mechanical solidarity* and *organic solidarity* to describe the types of relations between and among individuals in the two different societies. In the former, which emphasizes similarities among individuals, members share a common set of values and norms that are reinforced in routines and ritual. The degree of mechanical solidarity lessens as the division of labor progresses in the more modern society. In this society, organic solidarity emerges based on the mutual dependence of individuals as they engage in their specialized, but coordinated, activities. Individuals experience their duties and rights as the basis of their solidarity.

Two points are common to both Toennies's and Durkheim's theoretical perspectives. The first is that the transition from *gemeinschaft* or mechanical solidarity to *gesellschaft* or organic solidarity involves a change in the location of society from a rural or agrarian locus to an urban setting. This change is accompanied by increased population density. The second point is that the nature of shared values in the normative order changes from those that support the common activities of agriculture to those derived from an interdependence that is fostered by the extensive division of labor. Relationships in this setting are based on formal and financial reciprocities. Urban life is characterized by what Toennies calls "a condition of tension [of one] against all others." In his analysis of suicide, Durkheim claims that life in the society characterized by organic solidarity leads to *anomie*, a chaotic state of normlessness in which individuals do not feel compelled to abide by the prevailing value system.

This concept of *anomie* is similar to what Karl Marx (1818—1883) called *alienation* in his analysis of the relationship between the workers and owners of the means of production. Georg Simmel, the German sociologist, also spoke of this issue when he described the blasé attitude that develops in the money-based economy of the modern urban metropolis. All these theoretical perspectives point to changes in the normative order that accompany the shift from a rural, agrarian society to the modern urban form of social organization that continues to expand today. The increasing proliferation of ICTs suggests that the capability for establishing and maintaining new relationships between and among members of society may provide the opportunity to recreate some of the characteristics of the traditional or mechanical social structure. There are reports of participation in new types of social groupings organized around common interests or simulated societies. Many individuals have testified to the sense of well-being and competence they derive from interacting with others in online social networks such as Facebook or Twitter. This raises the possibility that many of the negative features of the anomic society could be ameliorated with participation in these new communities via ICTs.

LITERACY

Early Literacy

Many historians and social theorists point to literacy as a major transformative factor in societal evolution. Literacy allows for the intergenerational transmission of culture and thus the accumulation of knowledge, a major step in the creation of civilization. Of course, literacy did not suddenly appear in society. Reading and writing gradually spread throughout different societies over several millennia beginning as early as 5,000 years ago. Although many societies today are approaching literacy rates of almost 100 percent, in numerous places on the globe, less than 25 percent of the population can read and write.

Some evidence suggests that the first forms of literacy appeared in Sumaria where clay tablets stored information about commercial exchanges. The development of technologies that facilitate the distribution of various forms of literature has enhanced this evolution over the millennia. These technologies include the codex book with bound pages, paper and ink, printing presses, telegraph, radio, film, television, recording machines, and most recently computer networks. Note that the last five items all came into existence within the past century, and computers appeared only within the past half century. Despite accurate perceptions that society is indeed information rich, the technologies that support ICTs are relatively new.

Furthermore, the rate of developing and applying new technologies is escalating at an increasing pace.

It is informative to examine some of the controversies that accompanied the various transitions in the development of literacy. In ancient Greece, heated debates occurred over the predicted negative consequences of writing and reading. According to some classical scholars, Socrates was one of the most outspoken Athenians to criticize writing. He claimed that written documents would diminish the effectiveness of face-to-face dialogues. The meaning of a written word would become frozen, and misunderstandings and miscommunications would proliferate. We hear many similar complaints today about email. Socrates and others worried that writing and reading would diminish human mental capacities, for the great epic poems would no longer be recited from memory. It was widely feared that both the heuristic value and consummate pleasures of human discourse would gradually dissipate.

It is clear today that the dire predictions of the skeptical Athenians did not come about. However, it is also clear that modern conceptions of discourse are very different from those used by the ancients. In the Golden Age of Greece, discourse meant a face-to-face dialogue between a teacher and a student or among students to discover philosophical or mathematical truths. Today, a discourse might be a teleconference among a group of supply chain providers to agree on a schedule of production and shipment. Alternately, memory to the ancients was the capability to recall and perform epic poems such as Homer's *The Iliad* and *The Odyssey*. Memory in the present day implies the capacity to recall short- and long-term cognitive activities frequently with the aid of some information storage and retrieval system. The meanings of *discourse*, *memory*, and many other concepts have been altered in the course of transition from a pre-literate to a literate society. Some scholars have claimed that patterns of cognition have been fundamentally altered in the transition from an oral to a literate culture. In addition, some cognitive scientists have speculated that similar changes may occur with the transition to an information culture in which icons may become a common means of communication.

The Printing Press

In the fifteenth century, the transition to a literate society received an enormous impetus with the invention of the movable type printing press, usually credited to Johannes Gutenberg. Although the Chinese had earlier used clay and wood blocks for printing, Gutenberg's press used metal blocks, and this made it possible to print many identical copies of a manuscript and thereby reduce substantially both the time and costs of duplication. Originally, the press was used primarily to reproduce copies of the Christian Bible. Other religious manuscripts were subsequently reproduced and sold widely, and eventually printing was used for the distribution of secular materials. Similar to the response to the introduction of writing, many active critics opposed the printing process. Initially, the process could not include the lavish illustrations that usually accompanied the hand-drawn and copied manuscripts; the belief was that the reader of printed copies would experience a diminished sense of piety. It was also argued that an error in setting the type could propagate misunderstandings and eventually lead to widespread heresies. It is not surprising that much of the criticism of early printing came from the monasteries, where, for hundreds of years, the monks had been engaged in the copying and drawing tasks.

Printing gradually became more widespread across Europe, and the mass production of books became a viable commercial enterprise. The objections to the technology of printing were overcome for the most part by the financial success of publishing enterprises. Reading materials in the form of books, pamphlets, and eventually weekly and then daily newspapers became available to large numbers of people, and the concept of *informed citizens* took on new dimensions. Yet, it is important to keep in mind with respect to writing and reading as well as printing that these transitions with respect to literacy were not a single event. Nor did they occur evenly within or across societies. Rather, they are more appropriately viewed as processes that occurred over a long period of time, literally thousands of years and with uneven social class and geographical dispersions. The seeds or precursors of such developments are usually discernable well in advance of the dates typically associated with the innovations. Furthermore, the consequences of these transitions are still unfolding today. It is important to keep this perspective in mind during this exploration of the various applications of ICT in contemporary society.

ICTs and Literacy

Scholars have recently pointed out that literacy transitions in society are currently undergoing another major development with an increase in the application of ICTs in the creation, distribution, storage, retrieval, and use of information. Indeed, some analysts are now proposing a three-phase progression from a pre-literate to a literate and then to an information society. Albert Borgmann, a professor of philosophy at the University of Montana, presented such an analysis and labeled the three stages as natural information, cultural information, and technical information. Borgmann's contention is that as society moves more fully into the technical information stage, it is in danger of losing contact with reality. For example, he fears that listening to recordings of theatrical and musical events will take the place of live performances, and in the near future audiences will not be capable of appreciating the difference.

James O'Donnell, a classics professor at the University of Pennsylvania, also recognizes the new era of ICTs and compares the transitions of oral to written and copied to printed texts. Furthermore, O'Donnell explores the implications for teaching and learning and speculates on the future of a liberal arts education and the changing structures of institutions of higher education. O'Donnell's work compels us to consider another dimension of literacy—the need for students to develop a different set of skills to be effective and efficient users of the information systems that ICTs afford. Recognition of this need has prompted the Educational Testing Service (ETS) to develop a standardized test, iSkills, mentioned in Chapter 1, to assess information and communication proficiencies of secondary school and college students, with the assumption that these skills will be needed for lifelong learning activities.

Naomi S. Baron, professor of linguistics at American University in Washington, D.C., has completed extensive studies of languages used by young people on the Internet and with mobile devices. Her work focuses on emerging forms of communication and how they may alter social interactions. She also explores the conflicting claims that the Internet increases social interactions on the one hand and fosters social isolation on the other. Her analyses are highly regarded in the social science research communities.

The transition from a pre-literate to a literate and then to an information-based society provides a convenient and provocative framework with which to analyze societal evolution. Not too many years ago, literacy referred exclusively to the reading and writing of prose. However, in recent years, the concept has expanded to include other fluencies, such as arithmetic operations and mathematical reasoning. That skill is sometimes referred to as *numeracy*, a word formed by combining parts of the words *numerical* and *literacy*. The term *quantitative literacy* is also frequently used to refer to the same skills. Yet a third type of literacy, *document literacy,* refers to the interpretation of information in tables, graphs, charts, and images. Today, there is a much broader and increasingly accepted concept of literacy—*information and communication literacy,* referring to those skills required to locate, access, interpret, and apply information; it includes prose, quantitative, and document literacies. The National Adult Literacy Survey, also developed by ETS, assesses those three domains.

Another dimension of literacy that appears to be influenced by the increasing use of ICTs is the style of written communication. Social software is enabling easier communications over ICTs. Emails, instant messages (IM), blogs, Twitter, Wikipedia, MySpace, Facebook, YouTube, Instagram, and Snapchat all are engendering new and different patterns of communication. With increasing frequency, these communications employ multimedia combining written text and audio and video materials. The style of these communications distinguishes them from what is traditionally used in correspondence, books, magazines, and newspapers. There is frequently little or no use of punctuation marks or capitalization. The exclusive use of lowercase letters is becoming so common among young people that many consider the practice to be the badge of youth.

The use of acronyms is also becoming more widespread. For example, "lol" is used to denote "laughing out loud" or "btw" for "by the way." Sometimes combinations of letters and numbers are used as in "y r 18" for "you are late." Much of this style is employed as a faster and more efficient manner of writing. Some of the motivation may be to flout the conventional rules of grammar and etiquette. Evidence of this style in communication is found in all types of social software, but it is probably most common in the use of IMs and Twitter, particularly among adolescents and young adults. Some communications are quite opaque, and only those initiated in the jargon and acronyms can decipher the message. Many young users of IMs employ these systems to maintain constant contact and know the whereabouts of their circle of friends. The acronyms make it difficult for the uninitiated, perhaps parents, teachers, or other adults, to decipher the messages.

A professor at a large, urban university recently reported experiencing the new literacy style. During the semester, all the students in his graduate seminar participated in an online discussion group between the weekly meetings of the class. Over the course of the semester, as the students became more comfortable communicating with the professor and other students, the postings to the class discussion board transformed from the more traditional scholarly style to the new and more casual style that characterizes online social software communication. The following academic year, the professor served on the doctoral dissertation committee of one of the students in the seminar of the previous year. His assignment on the committee was to read the manuscript and question the candidate on the contents in a public, oral defense. After the first reading of the draft, the professor objected strongly to the literary style. It was filled with acronyms, run-on sentences, missing punctuation, and lack of capital letters. It read much like an extended Twitter post, and it seemed to the professor that the student had used the literary style that was practiced on the seminar discussion board.

The professor advised the student to rewrite the dissertation and to follow the conventional rules for scholarly publications. The student made a good faith effort to revise according to the professor's request; however, when the dissertation was resubmitted, it still contained many of the elements of style found in social software writing. At first, the professor was irritated, for he did not want to spend additional time remarking all the necessary corrections on the entire manuscript. He felt strongly that doctoral candidates should have mastered the techniques of scholarly writing by the time they submitted their dissertations. He decided to reread the manuscript and note all the places where the writing was either ambiguous or simply unclear and then ask the candidate to rewrite only those sections. Much to his surprise, the professor found very few instances where the more casual social software style interfered with the clarity of the points being discussed. Upon reflection, the professor realized that he was assuming that the new style would not be precise enough for a doctoral dissertation. Putting aside his own biases as to the proper format, he had to admit that the text was certainly clear. One could not fairly reject the candidate on the basis of lack of clarity. The professor wondered whether he was witnessing the beginning of a change in the style of scholarly communications.

Languages and conceptions of literacy are constantly changing. This anecdote of the professor and his student's writing emphasizes that the increasing uses of ICTs are producing changes in many commonly accepted practices in literacy, and it seems reasonable to expect that the further proliferation of ICTs will escalate the rates and patterns of those changes. The graduate students of today will eventually become the editors of scholarly journals in their fields, and their conceptions of clarity and appropriate styles will predominate. Furthermore, some speculate that if the use of icons and videos in ICTs continues, most forms of communication will eventually employ various forms of multi-media. Recently, several new operating systems have been introduced in which users communicate with the computer simply by touching tiles or icons on the screen. Android, Apple's iOS, and Microsoft's Windows 8 all employ this mode of user interface. If this trend continues, the combination of multi-media and iconography may be a new and fourth stage in the evolution of literacy. However, the written word will not disappear. Writing did not replace verbal communication in ancient Greece, and printing did not replace written communication in the late Middle Ages. Multi-media will most likely further expand and amplify our capabilities and patterns of communication.

ECONOMY AND POLITY

The third and last of the three categories of societal evolution theories focuses on economic and political dimensions. These theoretical perspectives examine the transitions from agricultural to manufacturing to knowledge-based economies and their attendant political systems. These transition stages are sometimes referred to as pre-industrial, industrial, and post-industrial. A great deal of overlap exists between the theories in this category and the two discussed earlier. However, the emphasis on political economy introduces additional elements that will be useful in subsequent analyses. The writings of Karl Marx, Max Weber, and the more recent work of Daniel Bell are examined in this section.

Karl Marx

Marx is widely known as the father of revolutionary communism. However, his substantial work on the historical evolution of society as driven by economic dynamics is especially well-known among the scholarly community. Marx saw an ongoing conflict between those who owned the means of production and those who labored in the production processes. In the agricultural society, landowners exploited the serfs, slaves, and peasants. In the industrial society, the bourgeoisie, or capitalists and factory owners, exploited the proletariat, or the men, women, and all too frequently the children, who worked for low wages and long hours, usually in unsafe conditions. In Marx's view, the transition from an economy based primarily on agriculture to an industrial one simply replaced an older system of exploiting the laborers with a new one that exacerbated their deplorable working and living conditions.

Marx predicted that the proletariat would eventually rise up in a revolution and establish a new society in which collective ownership would be the organizing economic and political principles of social structure. The dictatorship of the proletariat would replace the capitalist-controlled society, and all property would be owned collectively. Gradually the government would wither away, and a classless state would emerge. Marx's theoretical writings have had enormous consequences all over the world. One could readily argue that the course of major political events throughout the twentieth century and well into the new millennium has been influenced by Marxism and its intellectual and ideological progeny.

However, as is well-known, no communist governments have as yet withered away. Those governments that ruled in Eastern Europe during the Cold War period were never truly communistic. Rather, they were more like state capitalistic regimes. As far as we know, no truly classless society has ever existed. However, some evidence suggests that in modern societies the proliferation of ICT applications, which are clearly the products of capitalistic economic and political systems, is creating greater class divisions. In Marxian terms, this means that the most economically disadvantaged of the proletariat are neither getting the education nor learning the skills to enter the bourgeois class and benefit financially from the ever-encompassing ICT systems. This growing gap in education and skills is frequently referred to as the *digital divide* and is discussed in subsequent chapters.

Today, Marxists point out that one aspect of the theory does persist, and that is the continuing tendency of the ruling classes to exploit the proletariat. Despite the replacement of the bourgeoisie with the expanded public ownership of many corporations, a great divide persists between the affluent and the most economically deprived members of modern society. Some of the pioneers in computing and networking expressed the hope that the new ICT industries might offer opportunities for the disadvantaged in modern society to advance financially. The Marxian perspective encourages us to look for class differences in access to and use of ICT systems.

Max Weber

Max Weber (1864–1920) was a German sociologist who wrote on a great many topics that profoundly influenced the development of the discipline.

The range of his sociological inquiries includes the philosophical foundation of the social sciences, methods of sociological research, economic and
political foundations of social structure, comparative studies of the major
religions of the world, the nature of authority, and the characteristics of
the modern bureaucracy. Perhaps Weber's best-known work is his study
that traced the origins of the values of the modern capitalistic culture to
the teachings of the Protestant churches, especially the tenets of Calvinism.
Although Weber does not place his analysis of what he identified as the
Protestant ethic and the expansion of capitalism within the framework of
societal evolution, examining this work is informative.

According to Weber, after the Protestant Reformation, the conspicuous
consumption of the Roman Catholic clergy and elite laity was replaced by
the rigid tenets of diligent hard labor and frugality. Particularly in Northern Europe, these values were adopted by increasing numbers of people in
the Protestant sects. Weber points out that a combination of hard work
and meager lifestyles eventually produced substantial wealth. Unlike the
wealth accumulated by landowners in the Middle Ages, which was used
to support lavish and frequently hedonistic lifestyles, the ascetic Protestant
sects developed an ideology that supported the accumulation of wealth but
strongly condemned anything but the expenditure of that wealth to support
what they defined as God's work.

In his analysis of the bases of legitimate authority in society, Weber
notes a historical transition of power based on traditional authority in
early agrarian and small-town societies to an authority based on rational
reasoning in modern urban societies. Weber also discusses a third kind of
authority, charisma, which is based upon the personal leadership characteristics of one exceptional person. Although charisma is less relevant to the
analysis of ICTs and social change, the history of computers is enriched by
characteristics and behaviors of those who founded the major corporations:
Thomas Watson at IBM, Steve Jobs at Apple, and Bill Gates at Microsoft.
However, for the purpose of this analysis, the transition from traditional
to rational authority is more germane. Traditional authority has much in
common with *gemeinschaft* and mechanical solidarity. It relies heavily on
adherence to the normative order of the past and is exercised by those whose
status reinforces that order. Traditional authority is rarely challenged, and
it is modified slowly. Rational authority, on the other hand, is based on the
widespread acceptance of the legality of the normative order and the right
of those in designated positions to exercise that authority. Rational legal
authority is the foundation of the bureaucratic organization.

Although Weber was not the first person to use the term *bureaucracy*, he
was one of the early sociologists to recognize that society was evolving from
an earlier stage in which authority derived from a traditional basis to a society in which rational authority in the form of the bureaucracy was becoming the prevalent form of social organization. Weber pointed out that the
bureaucratic form could be found in such widely diverse organizations as

government agencies, business corporations, educational institutions, religious organizations, hospitals, and military units. All these organizations are hierarchically structured, and rules and regulations determine behavior. Bureaucratic officials are appointed to positions based on their knowledge and qualifications, they are usually paid fixed salaries, they are loyal to their organizations, and they view their appointments as careers with promotions frequently based on seniority. Whereas Marx saw the future as belonging to the proletariat, Weber saw it as belonging to the bureaucrat.

Weber's analysis of bureaucracy is most illuminating, and the behaviors of the officials he described can be found in a wide variety of organizations today. Although Weber does indicate that bureaucrats are appointed to their positions based on their technical competence, he does not elaborate on the problem of specialized expertise and the nature of authority relationships. As the need for expertise increases in today's society as a result of more widespread applications of ICT systems, changes in the nature of relationships between supervisors and their subordinates in bureaucracies can be expected.

Daniel Bell

In what is perhaps the most prescient, recent global sociological analysis ever produced, the eminent scholar Daniel Bell recognized changes that had been occurring for several decades in U.S. economy. He pointed out the transformation from the predominant pattern of manufacturing goods that characterized the Industrial Revolution to a pattern of providing services. Originally published in 1973 and reissued with a new foreword in 1999, Bell's *The Coming of the Post-Industrial Society* is now a classic in the field. The title of the book aptly previews its leitmotif. Bell further noted that the service industries were becoming increasingly dependent on the emerging information technologies. The technical knowledge and competence noted in Weber's description of bureaucracy were becoming driving forces in the new economy.

The advances in scientific knowledge and technology applications that occurred during and after World War II placed a premium on knowledge and expertise. Such advances as the conversion of matter to energy, rapid information processing on digital computers, and more recent applications of bio-genetics have all opened vast new areas for economic development. Those with the requisite competencies have advanced to positions of power and influence in society. Research and development activities mushroomed, and the skills to carry out these activities were in great demand. Universities and research laboratories provided the training for these activities, and Bell saw these institutions replacing corporations as major sources of power and authority in society. Certainly research laboratories and think tanks are emerging as influential organizations, but whether universities will fill these roles in the future is by no means clear. Recurring financial crises and demands for accountability in higher education may diminish their status as brokers of power and authority, at least in the near future. In this new highly technical environment, Bell saw competence and expertise quickly becoming the bases for merit and reward. The relevance of Bell's analysis to the inquiry of ICT applications in society is obvious, and his perspective provides a more comprehensive view.

Bell's analyses of the post-industrial society have now been substantially extended by Professor Sandra Braman, Department of Communications, University of Wisconsin–Milwaukee. In some sense Braman's work is an expansion both of Bell's concept of the post-industrial society and Borgman's specification that society is moving into a technical information stage. Braman names this next stage of societal evolution the "informational state." She sees it as succeeding the bureaucratic welfare state. The processes of creating and using information form a new foundation of power and authority in society. Braman presents a thorough analysis of the implications of this new power structure across a wide range of economic, political, and legal domains. Included in her inquiries are such topics as the relationships among information, power, and public and private governance; individual and state identities; geopolitical, network, and informational borders; and the balance between freedom of speech and national security. In one sense the informational state, as described by Braman, provides the most comprehensive perspective for the historical and sociological analyses presented in this book.

Of course, these depictions of various perspectives on societal evolution are mere sketches of a huge literature on the history, philosophy, and sociology of Western Civilization. With the possible exception of the work of Bell and Braman, none of these theoretical perspectives are adequately comprehensive. A complete history would include them all and many more. Nevertheless, for the purposes of the explorations of the relationship between ICT applications and societal evolution, they do provide a provocative platform from which to make observations.

3 Normative Order: Part One

The remaining chapters follow a common format, first presenting a brief review of the relevant sociological thinking on the patterns of social change in one of the major institutions as previewed in Chapter 1. Each chapter then examines evidence as to whether and how information communication technologies (ICTs) are influencing those patterns of social change.

THE CONCEPT

This chapter and the next examine normative order, the most comprehensive of all social institutions. Normative order consists of the shared values that account for the vast uniformity of human behaviors. These widely shared values are incorporated in all the norms, mores, folkways, role descriptions, laws, rituals, and customs that guide behavior. Many aspects of the normative order are formally stipulated in legislation, judicial codes, standard operating procedures, job descriptions, and organizational rules and regulations. In the absence of normative order, social chaos would ensue. All societies strive to maintain order and avoid chaos. In his famous treatise *Leviathan*, the seventeenth-century Enlightenment philosopher Thomas Hobbes described that chaotic state as one in which life would be "solitary, poor, nasty, brutish and short." Normative order, then, is a social structure that precludes social chaos. A glimpse of social chaotic conditions is provided by the looting, rioting, and other forms of lawless behavior that often follows natural disasters such as earthquakes, tsunamis, and hurricanes. Under these conditions, normative order frequently breaks down. Similar conditions of social chaos may occur during times of civil strife, war, and other conflicts. In the aftermath of the English Civil Wars during the mid-seventeenth century that resulted in the beheading of Charles I, Hobbes wrote *Leviathan*, arguing for the restoration of the monarchy to maintain social and political stability.

Adhering to the normative order frequently elicits positive reactions from others, and deviations generate negative sanctions. Most of the time, people abide by the normative order; that is, they dutifully complete their assigned jobs, pay their taxes in a timely and accurate manner, do not violate contracts, stop at red lights, and do not shout "fire" in a crowded theater. A great deal of human behavior is based on the expectation that others will adhere to the normative order. For example, in most nations, it is routinely expected that all motor vehicles will travel on the same side of the road and not exceed the posted speed limit. However, this norm is frequently violated. When speeding occurs and is detected, a citation may be issued, a fine paid, and automobile insurance rates increased. On the other hand, motor vehicles in a race on a legal track that is used solely for that purpose are exempt from any speed limits. Thus, some norms are relevant in many but not all situations.

The Processes of Socialization

Children learn normative order through the processes of socialization. At birth, all humans are capable of a vast array of behaviors. Yet over the course of a person's life, only small subsets of those potential acts are ever executed. In the process of growing from an infant to a fully independent adult member of society, each person becomes aware of and, in most instances, abides by the prevailing normative order. Through mimicking or modeling the behavior of significant others, children learn and become socialized to what most other people recognize as appropriate behaviors. This process of socialization involves learning both appropriate and inappropriate behaviors. Most behaviors of children produce either positive or negative sanctions. When children learn to say "thank you," they might receive a positive sanction, perhaps warm praise. When they throw a temper tantrum, they are likely to elicit a negative sanction, perhaps a "time out." Although many parents become exasperated that the socialization process takes so long, most children eventually incorporate courtesy and frustration management into their behavioral repertoire and learn what sanctions they can expect. The effect of ICTs on the processes of socialization is addressed in more detail in Chapter 5.

The process of socialization continues throughout life. When adults enter a new work place, they must "learn the ropes." In order to function effectively in the new environment, they must learn about the organization, their role in it, and how that role contributes to getting work done in that setting. For example, in many work settings, an employee is expected to solve most problems without assistance and to consult with supervisors only on the more difficult or important issues. Violation of this norm might bring about negative sanctions, such as the ire of the supervisor or a silent treatment from peers. When the violation of such norms by a new hire occurs less frequently over time, then the process of socializing that person into that aspect of the normative order of the work group is nearing completion. On the other hand, if the individual continues to violate the norm, that person may be deemed unsuitable for that particular job.

Similarly, when a family moves into a new neighborhood, the adults and children may be expected to learn and adapt to the local customs. It may be tacitly agreed upon by everyone in the neighborhood that all residents will maintain their lawns properly, or that there will be no loud noises after 10 p.m. on weekday nights, or that children will not ride their bicycles on the streets after dark. Such customs may never have been articulated or even openly discussed, but they may have emerged gradually over a number of years among the residents. If the new neighbors violate such norms, the older residents, through a variety of at first subtle and then increasingly direct means, may seek to constrain those offensive behaviors. If no strong negative sanctions are forthcoming from the other neighbors, that aspect of the normative order may wither away. However, if the new family has moved from another state, they may have to adhere to more formal aspects of the normative order, such as administrative laws concerning motor vehicle registration or inspections, or serious negative sanctions might follow. These examples demonstrate that norms may not be formally specified, but violations may still result in negative sanctions.

Origin of, Changes in, and Deviance From Normative Order

The origin of normative order is a complex issue. Observations made by scientists who study animal behavior consistently report patterns of cooperation. These patterns have been observed across a broad range of species from insects to primates. Among some primates, these patterns are quite extensive. Although there are no known historical records to substantiate the claim, it seems reasonable to assume that some rudimentary forms of normative order came into existence when early human nomadic tribes engaged in hunting game, gathering food, and caring for children. In order to carry out such tasks, there had to be some agreements on how to cooperate. Obligations and reciprocity among the participants were necessary to ensure continued existence. As human groups transitioned from hunting and gathering to agriculture, the sets of obligations and reciprocities became more complex. Undoubtedly, normative order evolved from such early beginnings, and today it is a vast array of complex values, customs, rituals, mores, folkways, and formal legal structures that both prescribe and proscribe behaviors. Many dimensions of normative order are formally specified in civil and common law. Furthermore, many of these same dimensions of normative order appear as tenets of various religious faiths, giving rise to both the distinction between and some overlap among ecclesiastical canon and secular law.

Normative order, it should be pointed out, is not a static phenomenon. Values, norms, and laws are constantly changing. When rapid changes occur in society, however, adjustments to normative order frequently lag behind. Sociologists describe such conditions as *cultural lag*. For example, it took decades to develop the legislation and enforcement systems for motor vehicle operation for the rapidly expanding use of automobiles and trucks during the twentieth century. Motor vehicle codes had to be created to specify appropriate and illegal behaviors on public roads; inspection systems had to be established to ensure that vehicles were properly maintained; and driver education and licensing programs were needed to determine that operators were competent. In addition, other innovations such as automobile insurance, car dealerships, gasoline stations, repair shops, trained mechanics, paved highways, and traffic control signage and lighting all added further complexity to the normative order surrounding the automobile industry and the use of motor vehicles on public property. The current dispute between the music recording industry and advocates of unfettered access to music files on the Internet is a prime example of a conflict caused by a cultural lag. Explorations in this book point to numerous instances when the escalating rate of change and utilization of ICTs creates such cultural lags.

Thus far, this discussion seems to imply that a core set of common values underlies the normative order and that there is uniform and universal acceptance of these values by all members of society. This implication must be quickly corrected, for it is a gross oversimplification and obviously fallacious. Just a cursory review of the daily news demonstrates that many human behaviors violate various dimensions of the normative order; these behaviors include fraud, violence toward others, and substance abuse. Yet, given the potential for violations of normative order in each and every person, it is truly amazing that there is not more such deviant behavior. The small amount of deviance is testimony to the effectiveness of the socialization processes. Even so, that small amount of deviant behavior usually attracts a large amount of public attention. Conceptually, deviance is the antithesis of normative order.

Violations of laws are considered crimes against society, and the legal system usually responds to detected crimes. There is a great deal of interest in the yearly reports of the occurrence of various crimes, such as murder, robbery, fraud, rape, and substance abuse. Increases in crime rates are frequently considered as indicators of the decline of the moral character of a society. Many pundits and social scientists offer hypotheses to explain the patterns and predict future deviance rates. Proposed explanations frequently include the decline of the nuclear family, increasing numbers of working mothers, rising environmental pollution, and falling standardized test scores. Indeed, analyses of these data never produce satisfying results, for the causes of deviance and crime are many, varied, and interrelated. Furthermore, official crime statistics frequently include only a small proportion of actual incidents.

Many criminal acts are never detected or reported and hence never recorded. Data on the incidence of rape are a prime example. Therefore, the changes observed in yearly crime reports might reflect in large part the variations in observation and reporting. Some social scientists have speculated that there is a constant rate of occurrence of deviant acts in any society, and all that varies are the rates of observation, recording, and responding. Furthermore, the definition of what is a violation of normative order may vary from one society to another or even at different times in the same society. For example, it was once illegal to produce and consume alcoholic beverages in the U.S. To this day, many people still consider drinking alcohol to be unwise, if not immoral.

Many social theorists, including Emile Durkheim and the contemporary sociologist Kai Erikson, have claimed that a certain amount of deviance is advantageous for a society, because it is the source of many innovations that subsequently become part of the normative order. In other words, deviance may contain the seeds of future social change. Innovations in the world of fine art, such as impressionism or cubism, were initially greeted with derision and claims of debauchery. The first public performance of the ballet "The Rite of Spring," with music by Igor Stravinsky and choreography by Vaslav Nijinsky in Paris in 1913, was followed by acts of violence against performers and even destruction of the performance hall by an outraged audience. The police were called in to quell the storm. Yet in a few short years, those same works of art were hailed as pioneering efforts that expanded the scope of creative expression.

Role of ICTs

Normative order provides another lens for examining the global conflicts raging today. Many of the values that underlie the normative order of the major religions of the world are identical, but there are many important differences, particularly when interpreted by individuals who take extremist views concerning such issues as holy wars or the roles of women in society. Since the destruction of the twin towers of the World Trade Center in New York City on 9/11, the differences in the normative order in North America and Europe on the one hand and some nations in the Middle East on the other have become painfully, and belatedly, more obvious. As ICTs enable greater communication among the populations of different nations, religions, and cultures, conflicts between and adaptations to various aspects of the normative orders in all countries should be expected. Such gradual and frequently conflict-laden adaptations are important components of the processes of globalization, and they are substantially facilitated by the proliferation of ICTs, including the more widespread distribution and viewing of Western-produced films and TV programs among residents in Middle Eastern nations. The contrasts in the lifestyles in Western and developing nations further add to the resentment that many feel as a consequence of the exploitation of those developing nations by what are perceived as former colonial empires. This is a new and growing form of cultural lag, and ICTs are increasing their salience in international affairs.

From this brief discussion of normative order, several simple conclusions may be drawn. First, it does account for a great deal of uniform and predictable behaviors most of the time for most people. Second, all members of any society rarely accept all the values inherent in its normative order. Therefore, at any given time a certain amount of deviant behavior can be expected. Third, in some instances, deviance may be a harbinger of a future change in the normative order. Fourth, normative order is never completely stable. Rather, it is in constant flux, influenced by such factors as migration, economic conditions, political alliances, and ICTs. Finally, conflicts in normative order frequently occur between and among different societies and nations. This last point does not provide an optimistic view for resolving current international disputes in the near future, for changes in the basic values that underlie normative order rarely happen quickly. Rather, they occur slowly as different values and behaviors become more widely accepted.

Normative order is probably the most comprehensive theoretical concept in all the social sciences. Its scope is so broad that it encompasses everything from the core values of social order to the contractual obligations among individuals and organizations to the specifics of decorum and etiquette and to interpersonal relations in small groups such as families, and dyadic relations such as doctor–patient, teacher–student, government official–citizen, and producer–consumer. The discussion of the changes in normative order that are related to the increasing proliferation of ICTs is also broad and comprehensive. To provide a structure to this discussion, the following six dimensions of normative order are discussed in the remainder of this and the next chapter: social inequalities related to the division of labor, expanded political processes, altered conceptions of intellectual and artistic ownership, definitions of individual and organizational privacy, new patterns of religious activities, and changing forms of criminal and deviant behaviors. The first two are examined in this chapter, the remaining four in the next chapter.

DIVISION OF LABOR

The first theoretical perspective of normative order to be discussed in the context of the effect of ICTs is the division of labor. It is one of the oldest and most frequently cited and recognized of all sociological concepts. In 1793, in *An Inquiry into the Nature and Causes of the Wealth of Nations*, Adam Smith described the natural tendency of humans to specialize in a task and then barter and exchange with others to obtain the results of their specialized labors. From their earliest beginnings, human societies have divided and shared those activities necessary to sustain life, such as providing food and shelter and protecting children. One person might specialize in making spears and knives for hunting and receive food from others in exchange. Or a miller might grind grains in exchange for other foodstuffs. The name of the concept aptly describes its essence.

By the beginning of the twenty-first century, the division of labor had evolved into a vast and intricate assignment of tasks to accomplish a wide variety of goals in all kinds of social organizations, including but not limited to families, schools, churches, civic groups, factories, businesses, governments, and international organizations. Indeed, the division of labor is the essence of the modern society. The members of any social collectivity are assigned tasks, and supervisors coordinate the results of these activities to ensure the accomplishment of common objectives and goals. The specific activities are described in role definitions, and the persons who fill the positions supposedly possess the necessary knowledge, skills, and competencies to perform effectively. A modern-day automobile production assembly line is an excellent example of the division of labor. Roles define the various activities along the production line, and supervisors ensure that the activities are coordinated to result in completely assembled and inspected automobiles or trucks at the end. Increasingly, computer-controlled robots or ICTs are completing many steps along the assembly line.

It is illustrative to revisit the previous discussion that hypothesized the emergence of normative order in early farming societies. The artifacts found in archaeological digs at the sites of some of these early societies suggest that the division of labor was based on the distinction between farming and home-making activities. The general pattern was that the males performed the former, and females performed the latter. Of course, there were many exceptions, but for the most part, this was the division of labor for many early human societies. Indeed, that general pattern has persisted over several millennia and still exists in a similar form in many parts of the world today. Division of labor, then, refers to the assignment and completion of different tasks that are integrated and coordinated to ensure the attainment of the goals of the collectivity. In early times, the goal was to sustain the farming community. In modern society, the division of labor occurs in all organizations, both private and public. Furthermore, this division of labor creates a hierarchy in which some people carry out activities to enhance goal accomplishment and others function as supervisors or managers of those activities. As with normative order more generally, the division of labor is constantly changing. Indeed, in some instances it changes rapidly; frequently, the introduction of ICTs to enhance goal accomplishment promotes such rapid changes. Many factors account for changes in the division of labor, but ICTs are becoming one of the major reasons for the frequent restructuring of tasks and hierarchies.

Gender and Age

As background for this discussion, it is useful to examine some examples of changes in the division of labor resulting from the introduction of new technologies. In early eighteenth-century England, women used spinning wheels and weaving looms to produce woolen cloth in their homes. This production process has been referred to as a *cottage industry*. While the men were engaged in other activities to support their families, most typically farming, the women were spinning woolen yarn and weaving cloth. Presumably, the women first made the cloth for their own family consumption; as their productivity increased, sales were made to other families and eventually to vendors. Production was distributed among the cottages. Many women who engaged in this work became quite skilled and produced high-quality cloths, frequently with exquisite artistic patterns. In the cottage industry era women managed the necessary functions and oversaw the processes of producing the wool and weaving the cloth. Nevertheless, it was slow, tedious, and time-consuming work, and the returns on those investments were relatively low. This division of labor in the production of cloth was stable for several centuries.

As the Industrial Revolution swept through England in the late eighteenth and early nineteenth centuries, water mills and steam-driven machines supplied power for a variety of functions, including centralized spinning and weaving factories. These more efficient and productive factories gradually replaced the cottage industry. When these functions were taken over by the centralized factories, they became more complex and required a more elaborate division of labor, including operating machines, setting the tolerances for the varying grades of wool and cloth, feeding the raw materials, stacking and packing the finished products, monitoring the quality of the processes, servicing the moving parts of the machines, and even cleaning the floors around and under the machines to prevent a buildup of residues from the mechanized looms. Typically, women and older children were the operators, men set the machines and supervised the processes, and small children kept the area clean. Already in the earliest woolen mills, a gender- and age-structured division of labor was established. Gradually, this type of division of labor spread to other industries, including gun manufacturing and machine production.

When Henry Ford developed his assembly line to mass-produce automobiles in the U.S. in the early twentieth century, he created the most complex division of labor the world had seen to that date. Each step in the process of manufacturing automobiles was positioned on the ever-moving assembly line. The line started out with just the chassis. As it progressed, each component was added, and a completed automobile rolled out at the end of the line. Each job along the line had a unique and specific function, such as installing the engine; inserting the seats; mounting the doors; placing the hood, fenders, and running boards; and attaching the tires. At various points along the line and at the end, quality control checks were made. Ford's assembly line was enormously successful, and he was able to mass-produce so many automobiles with low sales prices that they were affordable by the employees of his own assembly lines. In the early days of automobile production, the assembly line was staffed with men, and for the most part they were younger men.

As the industry matured, the employees included older men and increasing numbers of women. The shortage of males for the assembly line production of military vehicles for Word War II substantially increased the ranks of women in the factories. Both the demands of the assembly line process and the availability of employees helped create a new age and gender structure in the division of labor in the automobile industry. For the most part, Ford's assembly line mode of production is still employed in the automobile industry, but computer-controlled robots accomplish many functions along the line today. The assembly line model is now employed in most manufacturing industries worldwide. Perhaps the most dramatic instances today of the integration of ICTs and the division of labor are the assembly line that mass-produces computers. Ironically, computer-controlled robots mostly drive the production. As more ICTs are applied to production processes, expertise is introduced as another dimension of change to the division of labor.

Expertise

Expertise is an especially interesting aspect of the division of labor in modern society. The salience of expertise is illustrated by examining comparative examples of military organizations. Consider first a U.S. Army infantry company at the time of World War II. The weapon most commonly used at that time by infantrymen was the M1 rifle. It was a fairly simple piece of equipment to use and to maintain. Every member of an infantry company knew in detail how to fire the rifle, as well as how to take it apart, clean it, and reassemble it quickly for immediate use. In fact, a great deal of time was spent practicing both firing and maintaining the M1. If the company commander observed a soldier cleaning his rifle and reassembling it incorrectly, he could reprimand him on the spot and show him how to do it properly. The commander was as competent in those procedures as was everyone else in the company. Expertise in the maintenance and use of the M1 was distributed equally. Experience in the use of the rifle was not equivalent, but the expert knowledge was. Furthermore, everyone in the company recognized that everyone else was equally knowledgeable.

For comparison, consider a U.S. Army infantry company today. The soldiers now use an M4 or M16 rifle, and most of these rifles are now equipped with an advanced combat optical gunsight (ACOG). This device uses a fiber optic system to collect available light to improve acuity and contrast for both daytime and nighttime operations. The devices are expensive and are not distributed to every company. Even so, few members of an infantry company today have the expertise to maintain or repair the ACOG system, and it is widely recognized that commanders, other officers, and NCOs are not equally competent with this technology. If the ACOG malfunctions, support services are either scarce or unavailable. All members of the company recognize that experience and expertise in ACOG are lacking. The nature of the authority relationships between and among members of this company is qualitatively different, because expertise is not uniformly distributed throughout the hierarchy.

Automation

Patterns of change in the division of labor resulting from ICTs are emerging in many organizations. Numerous blue-collar manufacturing jobs and white-collar clerical jobs have been taken over by automation. For example, telephone switchboard operators are now almost extinct, having been replaced by automated switching systems. Furthermore, many service industry functions are either accomplished by ICT systems or outsourced to other parts of the world. Telephone calls to customer service offices and help lines used to be staffed with people. Increasingly, those calls are routed to automated response systems equipped with pre-recorded answers to the most common queries. Alternately, the calls are answered by staff in distant countries earning much lower wages than those paid in the U.S. ICT networks enable such outsourcing. These trends are contributing to growing unemployment rates in developed nations. At the same time employment opportunities are increasing for jobs requiring expertise in ICT systems, but there is a shortage of qualified people to fill those positions. These trends have clear and obvious implications for future education and training programs.

An interesting automation innovation is occurring in the fast food industry. Increasingly when customers in North America place their orders for takeout food at a drive-in restaurant window, they are actually speaking to an operator in a country in the Far East. The order is then sent by an ICT network back to the restaurant where it originated for fulfillment and delivery to the customer. It is cheaper for the fast food industry to automate and centralize order receipt in a country where labor is cheaper.

Race

The division of labor has traditionally resulted in many social inequalities reflected in the dimensions of race, gender, and age. Until quite recently, the division of labor was so structured that almost all the roles in society that were accorded power, authority, and prestige were occupied by middle-aged white males. In the latter half of the twentieth century, a number of changes took place that began to reduce the inequalities along these dimensions. More women began participating in education programs and in the workplace. Although institutional barriers to gender equality still exist, much progress has been made in increasing the number of women in higher education and in influential positions in both business and government. Although they are controversial and frequently challenged in the legal system, affirmative action programs for admission to educational opportunities and color-blind selection processes have increased racial diversity in many organizations. And these legal precedents have encouraged older people to seek protection from age-related biases.

For many generations, race has been a major factor in both the division of labor and social stratification. In North America and parts of Europe where racial and ethnic diversity are common, a relatively stable, two-tiered society has emerged, and most people can be categorized as belonging to the "haves" or the "have-nots." As noted in Chapter 2, this is also referred to as the digital divide, and a lack of expertise in ICTs is an important contributing factor. Many social scientists acknowledge the increasing importance of specialized knowledge or expertise to individual economic well-being, as well as goal accomplishment in organizations. By examining the accumulation and distribution of different kinds of expertise in ICTs among individuals and in organizations, some insight may be gained as to future forms of the division of labor.

Social Inequalities

Although much progress has been made in eliminating inequalities based on gender, race, and age, vast inequities still exist in equal access for all members of society. However, the proliferation of ICTs, particularly in the workplace, has the potential of reintroducing inequalities in the division of labor. Women have traditionally chosen educational programs and careers that do not involve science and technology. Although more women than men are now enrolled in higher education institutions in the U.S., most of these women are not majoring in science, mathematics, engineering, or technology fields. The Association for Computing Machinery, the world's largest educational and scientific computing association, is desperately seeking to entice more young women to pursue careers in computation. A number of theories attempt to explain why more women do not enter science and technology careers, but thus far none of the strategies intended to ameliorate the dearth of females in these fields has yielded significant improvements.

It is also true that some minority groups, particularly African Americans and Hispanics, have not been attracted to technical fields. In addition, age inequalities in the division of labor are likely to persist for some time, because technical facility and competence in the use of the ICTs is much greater among younger people. It is possible that as today's young people mature and seniors leave the job market, the gap in technical competence between the young and the old may diminish. On the other hand, the rate of change and innovation in technology seems to be escalating, and the gap between today's young and tomorrow's young may persist. Furthermore, inequalities may develop along a new dimension. Immigration, particularly from developing nations, is introducing into highly technical countries large numbers of people who are not skilled in technology. These immigrants are seeking places in the labor force. For the most part, they have not yet acquired the technical competence or the language skills to qualify for many positions. In this case, ICTs may introduce a new dimension for creating and sustaining inequalities in the division of labor.

The division of labor is undoubtedly the most ubiquitous social structure in modern society. It is found in every social grouping, and all people have multiple positions in the divisions of labor among the many different groups to which they may belong, including family, workplace, religious organization, and political party. Indeed, one of the challenges of modern life is to shift effortlessly and efficiently from a specific location in the division of labor in one social setting to a different collection of responsibilities and authorities in another. The following chapters look for instances where ICTs are affecting the division of labor in various social institutions.

POLITICAL PROCESSES

The processes that create and sustain a governing structure to support the commonly accepted normative order are the foundation of every society. For the purpose of this study of the effect of ICTs on social change, the discussion of political processes is focused primarily on those activities that members of a society engage in to sustain a civil government that supports their prevailing normative order. It is true that political processes take place in all forms of nongovernmental organizations as well. However, the initial focus of this analysis is on governments. Some refer to these ICT-enabled processes as *egovernment.*

In democratic governments, the political processes are most extensive and complex, and public elections are the keystone of the system. Through elections, individuals occupy positions in which they enact and enforce the laws and common procedures that formally implement the values of the normative order. However, the activities of the officials and those they appoint to public management positions are frequently influenced by the preferences of those people who have contributed both funds and activities to their election campaigns in the past and will likely assist them in keeping their positions in the future. This assistance is provided primarily in the form of money used to pay for what appears to be ever-expanding expenditures for political campaigns. Therefore, the implementation of the normative order commonly reflects the values of those who provide that financial support. This is usually referred to as *lobbying*, and despite considerable controversy and calls for reform, it has become a common practice at all levels in most governments.

Voting

ICTs are exerting considerable influence in three major areas of political processes: voting, service delivery, and political action groups. The applications of ICTs in all three areas are causing changes in operating procedures and are frequently spawning vigorously debated conflicts. Perhaps the most controversial is electronic voting. In the early days of networked computing, there was much discussion and naïve enthusiasm that ICTs might enable continuous electronic plebiscites. The first advocates were enthralled by the possibility of strengthening the democratic foundations of government by making it possible for large numbers of people to participate conveniently and frequently in online political processes.

These enthusiasts talked of recreating what they perceived to be the political structure of the ancient Greek city-state governments in which all citizens participated in the legislative processes. However, such a utopia of universal democracy never really existed. It is well-documented that all citizens in many of the ancient cities and provinces were entitled to participate in the legislative processes, but not all did so at the same level of activity. Furthermore, the citizens made up only a small proportion of the total residents of these areas. The largest groups were slaves, and they were entitled to very few, if any, rights, and never participated in the legislative processes. Many social and political theorists and practitioners have realized that universal and continuous voting or referenda would rapidly lead to political confusion, and perhaps even chaos. It might be analogous to enacting legislation based on the results of a weekly public opinion poll. Upon reflection, most support for the concept of ICT plebiscites rapidly evaporated.

More recently, and realistically, public attention has shifted to the issue of electronic voting, meaning that some portion of the processes of registering voters, casting ballots, accumulating tallies, and transmitting results would involve computers and networks. As with plebiscites, the initial reactions to the possibilities of electronic voting were positive and enthusiastic. However, here as well, as more experience accumulated with online voting, security issues proved to be very complicated. At best, electronic voting systems are vulnerable to technical bugs that can distort tallies. At worst, most such systems are amenable to manipulation and almost invite attempts at election fraud. Numerous investigations of online voting systems have demonstrated how easily they can be breached to deny access to legitimate voters, illegally exclude large blocks of ballots, and ultimately produce incorrect and biased tallies.

Bills have been introduced in federal, state, and local legislatures to require that a paper record of all votes cast be produced by electronic voting systems. However, these bills have been stalled by opponents who claim that the costs to initiate this reform would be prohibitive. It is difficult to discern the motivations behind those who oppose paper trails. Cost does seem to be a legitimate concern. However, a long history of various modes of election fraud in many countries might lead one to be suspicious of attempts to thwart electoral reforms. Paper trails could go a long way to increase the integrity of electronic voting. However, in addition to their expense, such systems do not provide ironclad protection against breaking into the system and distorting election outcomes. Tests conducted on commercial systems that have been widely used by numerous states reveal the need for hardware and software modifications and close monitoring of the systems before, during, after, and between elections to ensure accurate vote counting.

Some comments about online elections in private corporations and organizations are relevant, for the use of ICTs in voting is not limited to public agencies. In fact, their use in both public and private organizations is increasing, and charges of abuse and fraud pose serious threats to normative order. Public corporations are required to publish quarterly and annual reports on their activities. These reports include the results of their shareholders' votes at annual meetings for directors, officers, auditing firms, and corporate policies or practices. Shareholders have traditionally been invited to attend these annual meetings and to vote on all the issues. However, only a few actually attend, and management casts the proxy votes of those not in attendance at the meetings. The processing of paper ballots is expensive and cumbersome, and shareholder participation has always been minimal. Today almost all major corporations circulate their ballots and collect proxy votes using ICTs. It is not yet clear whether ICTs are producing greater or diminished shareholder participation in corporate management. However, the potential for change in the governance of publicly owned companies is, for the most part, yet latent and unrealized.

Many professional and voluntary associations increasingly use online voting systems to elect officers or amend their constitutions. To date such online voting systems have generated little controversy. Elections in such organizations are rarely contested, but when controversy arises, the accuracy and validity of the online election system are questioned. For example, college and university alumni associations traditionally solicit nominations and votes by sending paper ballots through the mail. Many of these organizations now offer their members an optional online ballot. They report that online voting is growing in popularity, and paper ballots may disappear completely in the not too distant future.

A group of younger alumni of a well-known private college recently mounted an effort to exert more influence on the election of alumni representatives to the board of trustees. The college administration had played a major role in selecting all new board candidates. The younger alumni wanted to introduce additional and external perspectives, and they questioned the transparency and integrity of the online voting system. The allegation of a lack of integrity resonated with many of the alumni, for the operation of the online system was completely in the hands of the college administration. Although ICTs increased the visibility of these processes for selecting trustees, the procedures for soliciting and tallying votes could be readily observed only by an expert in the hardware and software of the system. In this case, employment of the ICT facilitated the generation of a new conflict between the young alumni and the college administration.

The increasing use of ICTs in both public electronic voting and private organization balloting systems presents a duality of opportunities. On the one hand, ICTs can increase certain transparencies, comprehensiveness, and efficiencies. However, on the other, it invites potential manipulation, duplicity, and fraud. Such attributes were equally inherent in the manual or paper voting systems in use prior to the advent of ICTs, but the technology permits new opportunities for deception, or even corruption. If the latter does occur, expertise will be required to discover it; but that expertise will not be widespread in society. Most people will not be able to make that judgment, and widespread distrust may become common. Even if corruption never occurs, the confidence of many people in both private and public electoral systems will be challenged. This issue of positive and negative potential impacts of increasingly widespread use of ICTs is a theme that is common in the exploration of normative order, and indeed of all major social institutions.

Service Delivery

The provision of services by government agencies is a second critical component of the political processes that help maintain the normative order. The number of government agencies in many countries that are using ICTs to enable procedures and communications is increasing. As mentioned earlier, these activities are commonly referred to as *egovernment*. In the U.S. the increases are occurring at all levels of government, including federal, state, and local agencies. Even American Indian tribal governments are employing ICTs for service delivery. However, the bulk of these activities are occurring at the federal level. Furthermore, this pattern is repeated in many other nations that are developing extensive ICT systems. In fact, the Republic of South Korea is currently the world leader in delivering services to citizens by government agencies that employ ICTs.

Three categories describe the services that are provided by ICT-facilitated exchanges. First are exchanges between government agencies and individual citizens; second are those exchanges between government agencies and private organizations, mostly businesses; and third are exchanges between and among government agencies. The comments here pertain to the U.S., but similar patterns exist in other nations. A growing use of an ICT-facilitated service between government agencies and individual citizens are the electronic filing and payment of individual federal, state, and local income and real estate taxes. Tax refunds are also increasingly processed using electronic funds transfer systems. The submission of online applications and issuances of a wide variety of licenses and registrations are also becoming widespread. These include motor vehicle registrations and drivers' licenses, professional and occupational licenses and certifications, passport applications and renewals, and even hunting and fishing licenses.

All these activities are now routinely accomplished online. Fees for such services are collected either with credit cards or direct withdrawals from individuals' bank accounts. In addition, the Social Security Administration accepts online applications for its services and payments, including updating of files. Furthermore, most payments of retirement and disability benefits are accomplished as electronic fund transfers. The delivery of police, fire, and first aid services is greatly facilitated through the use of ICTs for rapid dissemination of alerts and the location of individuals in need of service. Many federal, state, and local government agencies also post online employment opportunities for citizens to access and submit application forms.

Service deliveries between government agencies and private organizations, the second category, are also being effectively facilitated through the use of ICTs. Business organizations can comply with revenue regulations online and make tax payments, which include employees' withholdings, Social Security contributions, corporate tax payments, and mandatory financial reporting. Businesses also file online patent and trademark applications, articles of incorporation, and declarations of customs payments and exemptions. Although government-to-business exchanges do occur at state and local levels, the majority of such transactions are with the federal government.

The third category of ICT-facilitated services, activities between and among government agencies, is more limited than the first two. For the most part, they are confined to sharing of databases, and this is most common among intelligence and law enforcement agencies. For obvious security reasons, most of this activity is kept from public view, and estimates of its volume are difficult to obtain. However, as concerns for protecting individual privacy grow, demands for more information on the activities of various government agencies in data collection and sharing are increasing. Many protests point specifically to those agencies that gather data relevant to intelligence monitoring of potential terrorist activities.

The efficiency of such sharing of intelligence files among government agencies in the U.S. has come under scrutiny and increasing criticism recently. As the U.S. government began to realize that it was engaged in an ongoing war against terrorism, it increased its efforts to collect intelligence data on individuals and organizations that it suspected of posing threats. Such records had been kept by government agencies in handwritten files for many years, but the combination of the capabilities of ICTs and the growing perception of terrorist threats prompted new attempts to increase the size, scope, and coordination of such intelligence files across agencies. The recent case of an employee of a private firm under contract to the National Security Agency disclosing publicly such data gathering has caused a great deal of concern around the globe. It is not clear how this incident will impact these activities. At a minimum, public attention has been focused on the question of how much information the government should collect about the activities of citizens without initiating an open court-approved search warrant. Attention also has been called to the difficulties of gathering and sharing intelligence data among different agencies. For example, it remains practically difficult to create and utilize a list of people who should be prohibited from boarding commercial aircraft flights bound for North America. Efforts to share such data among agencies from different nations raise serious questions about common standards relevant to the propriety of such activities.

For the most part, online government service delivery is widely viewed as a positive development. Many people are spared the inefficiencies of standing in line in public offices to accomplish routine tasks. Agencies are able to operate with increased speed and efficiency. The ICTs facilitate the collection, analyses, and reporting of data on the nature and volume of services delivered. It is generally accepted that ICT-enabled service delivery saves taxpayers money and eases many bureaucratic burdens. However, one downside to ICT-facilitated service delivery that is infrequently discussed is that lower-income individuals and families have less access to and familiarity with ICTs. Consequently, they must still rely upon face-to-face interactions in government agency offices or exchanges using the postal service. All too frequently, lower-income individuals and families are unable to take advantage of many online services.

The use of public goods and services has always favored the more affluent. This differential is now well-documented and widely recognized. For example, many years ago pre-school children from middle- and upper-class families were shown to benefit more than those from lower income backgrounds from watching the PBS broadcasts of *Sesame Street*, a television series designed to help all pre-school children develop reading readiness skills. The parents in the more affluent families watched the broadcasts with their children, and they bought the books, recordings, and toys that reinforced the content of the programs. Consequently, the net effect of the program was to widen the reading performance gap between the children of the rich and the poor. Likewise, affluent persons and families are usually able to obtain superior levels of services from many other public or quasi-public facilities such as museums, libraries, hospitals, and schools. ICTs simply exacerbate this inequality.

Many urban governments are considering establishing citywide wireless zones so that all citizens will have continuous access to a variety of services via ICTs. Some state governments have announced intentions to support wireless access to ICTs in rural areas that include many people and families living below poverty levels. Given the declining costs and increasing power of network systems, it seems clear that wireless zones will eventually expand their coverage. In part, this will help address some of the issues of the digital divide. However, until lower income people have both access to ICTs and the necessary skills to use them, the problem will persist. The divide may even be expanding now, because the middle and upper classes are increasing both their access to and their skills in using ICTs.

The government in Hong Kong has recently announced its intention to create a program that will address the issue of the digital divide within its jurisdiction. Hong Kong is a most unusual city. From the middle of the nineteenth century until 1997, it was a British-controlled territory. It is now a special administrative region under the People's Republic of China. Its culture and economy reflect both Chinese and British influences. It is today one of the most important capitalistic trading centers of the world. Its population of 7 million people is crammed into less than 500 square miles. By comparison, New York City has over 8 million people in 468 square miles. The soaring skyscrapers of Hong Kong are recognized throughout the world as the epitome of modernity. It is reported that its extensive public transportation system accommodates more than 90 percent of its residents' daily travel. Hong Kong is a densely packed modern urban center with a vibrant free market economy.

The Hong Kong government estimates that 80 percent of its population uses the Internet on a daily basis. This rate of participation is much higher than that in most other nations. The rate for the U.S. is estimated to be somewhere between 40 percent and 60 percent. The Hong Kong government now plans to make available all the necessary hardware, software, training, and support to what it calls the last quintile, the other 20 percent of the population that is not yet using the Internet. The people in this last quintile are mostly elderly, disabled, low income, and less educated, precisely those who require the most extensive training and support to become effective Internet users. Many other national and local governments around the world have announced plans for expanding networks to reduce the digital divide, but none are as ambitious as Hong Kong, which intends to eliminate it completely.

Political Action Groups

The third area where ICTs influence political processes enables action groups to expand substantially their efforts and effectiveness in influencing the election outcomes and the behavior of government agencies and officials. Among the three branches of government, judicial, legislative, and executive, the latter two are particularly subject to the sustained efforts to influence by special interest parties. Indeed, many political theorists claim that this process of exerting pressure on government agencies and officials is a critical component of a viable, modern democratic system. Others claim that such a system of influencing legislators and executives makes it possible for the affluent to shape government policies and practices to their particular interests. ICTs, and specifically the Internet, make it possible to mobilize all sorts of political activities for a wide variety of objectives and causes. The list of organizations that make extensive use of the Internet to promote their agendas and perspectives is extensive and expanding rapidly.

Throughout the world, a veritable cornucopia of websites, blogs, and mailing lists now exists to promote the interests of all positions on the political spectrum, as well as countless special interest groups. Liberal groups in the U.S. include the Democratic National Party, Common Cause, MoveOn, American Civil Liberties Union, and the Civil Rights Project. With websites and online newsletters, these organizations promote the election of liberal members to Congress and state legislatures. Furthermore, they encourage their subscribers to contact their elected representatives and encourage them to support or oppose specific pieces of legislation. On the conservative side of the political spectrum are such organizations as the Republican National Party, Foreign Policy Research Institute, American Conservative Party, Cato Institute, Family Research Council, and the Ethics and Public Policy Center. They engage in similar ICT-based activities to promote conservative causes and candidates.

Hundreds more political organizations maintain a presence on the Web representing all positions on the continuum of political ideologies. In addition, countless organizations use ICTs to enlist support for their specific causes. These organizations solicit funds to support their activities and ask for signatures on petitions to a wide variety of public and private agencies to initiate new or modify existing policies or programs. Other organizations encourage contacting elected or appointed officials to promote their social, political, or financial agendas; such contacts are made quite readily using email. The activities of organizations on both sides of the recent debates over gun control and women's reproductive health issues were greatly intensified with the use of ICTs for raising awareness and funds. Prior to the emergence of ICTs, advocates had to write letters or make telephone calls—both time- and resource-consuming activities. However, current ICTs enable one to send many pleas to elected or appointed officials with just a few clicks on a Web page. However, it is not clear just how much elected or appointed officials are influenced by the number of emails they receive. Many proclaim that they want to hear from their constituents, but how much it may change their position or vote on any given issue remains unknown. Undoubtedly, the influence of barrages of emails will vary substantially depending on the specific issue. Perhaps a research project that asks elected officials and their staff members to respond confidentially about the impact of emails from constituents on voting behaviors might shed some light on this issue.

The activities of groups of political partisans have recently taken on a new dimension that is bringing about turmoil and dramatic changes in governments in the Middle East. Just how much the use of social networks such as Twitter and Facebook contributed to the extended and extensive political demonstrations and protests in Turkey, Tunisia, Egypt, Bahrain, Libya, and the Ukraine is still unclear. Reports from participants in the demonstrations and foreign journalists on the scenes claim that social networks were crucial to organizing and managing the political activities that brought down some long-established governments and threatened the stability of the entire region. The final assessment of the role played by the social networks will await future research. However, it does seem clear that social networks have the potential to cause rapid and dramatic political change.

In the next chapter, the analysis of normative order and ICTs continues with changing perspectives on the concept of ownership.

4 Normative Order: Part Two

OWNERSHIP

Ownership is another fundamental component of normative order. To maintain a stable society, there must be a near universal agreement that individuals and groups have the right of ownership that is protected by the norms, values, and laws of the society. According to a basic premise of any orderly society, one does not take what someone else owns. Three types of ownership exist. First, and originally, ownership refers to possession of a piece of land, such as a farm, residential property, or commercial site. Second, ownership refers to objects, such as animals, automobiles, furniture, or jewelry. Until recently, this category included the ownership of other human beings. Slavery was not abolished in the U.S. until late in the nineteenth century. In ancient history, slavery was widespread in many cultures. Unfortunately, its vestiges are still prominent in many nations today, and they sustain many of the inequalities discussed in the section on the division of labor in the previous chapter. Finally, ownership refers to intellectual products, such as a copyrighted book, a musical score, a piece of artwork, or a patented invention. Other than their use for maintaining records, information communication technologies (ICTs) have thus far had little impact on the first two types of ownership. However, ownership of intellectual products is being radically altered by ICTs.

Intellectual Products

The increase in the means and power of distribution enabled by ICTs creates many complex new problems in determining the ownership status of intellectual products. The difficulties arise in trying to establish the origins of and draw boundaries between and among such items as ideas, compositions, artworks, recordings, and motion pictures. And the difficulties are further compounded when ICTs are used to create new works that contain components of multiple, earlier copyrighted materials. For example, is it legal to produce, distribute, and sell a new motion picture that combines extracts of copyrighted materials from earlier films with images of paintings and tracks of music recordings? The answer to that question is by no means clear, and it is not likely that it will be resolved in the near future. Combinations of materials from different original sources are called *mashups*, and they are exceedingly easy to produce with ICTs that access such materials on the Internet. It is evident that ICTs are changing the ways that we are creating and distributing copies of new creative works such as music, film, and books. All these changes are leading to a new concept of ownership, often referred to as *open source* or *free culture*.

Most countries have developed a series of laws and procedures for protecting the ownership rights of the creators of intellectual products, and a number of international agreements accomplish similar protections across national borders. The rationale behind these systems is to provide incentives to increase the production of intellectual output. These legal mechanisms include both copyrights and patents. Copyrights date back to the seventeenth century when the use of printing presses spread throughout Europe. As noted earlier, initially, the presses were used for printing religious texts, and churches regulated the production, distribution, and sale of the output, as they had for centuries controlled the hand-copied manuscripts. Gradually, the use of printing presses for the production of secular materials increased, and in England printers formed an organization called the Worshipful Company of Stationers and Newspaper Makers. The aim of the stationers was to control the finances of this emerging business. Concerned that the primary motivation of the stationers was to maximize their profits, the British Parliament in 1709 passed the Statute of Anne, named after Queen Anne, the ruling monarch at the time. The official title of the statute was "An Act for the Encouragement of Learning, by vesting the Copies of Printed Books in the Authors or purchasers of such Copies." It is recognized as the initial copyright law. As the name implies, the statute was intended to promote the production of artistic and educational materials by giving the authors the exclusive right to sell copies of their work.

Most nations have passed copyright laws giving authors the exclusive right to their materials for periods ranging from seven years to eternity. Such laws fostered the growth of the publishing industry, which purchased the rights to print and distribute copies of books from the authors, and paid them a percentage of the sale price of each book. The author's fee became known as the *royalty*, in recognition of the Statute of Anne, a law passed by Parliament and approved by Queen Anne. Subsequently, patent laws were established to provide similar incentives and a compensation structure for inventors. Patent laws do not give rights to produce and sell inventions. Rather, they exclude others from using the invention for their own financial gain without the permission of the patent holder. Both copyright and patent violations are processed in civil courts.

Many legal scholars are beginning to claim that the system of copyright and patents has become obsolete. They were enacted over many years as creative incentives and protection of intellectual property in response to the production activities of the original Industrial Revolution. However, the economics of the present era in which ICTs are so commonly used requires drastic revision and update of copyright and patent laws. Consider the creation, production, distribution, and acquisition of books that are protected by copyright. Most book publishers use some system of computer-aided production, and a machine-readable, digital copy of the book is a byproduct of the printing. Since the cost of scanning a printed book is no longer exorbitant, digital copies of new books abound and are available for large-scale distribution and sale in violation of the copyright. No one knows for certain how large the business of pirated books is, but it seems widespread throughout the world. Some claim that such distribution is a boon to scholarship and cultural enhancement; others claim that it is illegal and should be pursued by law enforcement agencies and those convicted of pirating should be punished. The capability of copying and distributing books, as well as other media materials, is made possible by ICTs, and it is giving rise to an intense debate as to its propriety.

Google has begun a project to copy all the books in the collections of the world's largest academic and research libraries. At the start of the project, more than 30 such libraries had committed to the project, and many others' participation was being solicited. However, the legality of this project has been challenged by the book publishing industry, and it is being investigated by the U.S. Department of Justice. One of the many issues that complicate such a project is the fate of so-called *orphan works*. These are books and other materials whose copyright is still valid, but whose authors or copyright holders cannot be located. Scanning such works and distributing them would be a violation of the copyrights. Since the passage of the Copyright Act of 1976 in the U.S., orphan works have become a more complicated issue, because the act eliminated the need to register a copyright. Therefore, the status of many recent and older works is unknown, and many could be orphans. Including them in the Google project could be illegal. Google has proposed a compromise position that includes full-text scans only for materials with an expired copyright or that have been transferred to a publisher. For all protected materials only bibliographical data and brief excerpts would be included in the Google Book Project files.

An additional concern is that a huge number of older books in the stacks of many libraries are decomposing to powder because of their acid-containing paper and bindings. These volumes will be lost if they are not copied and deposited in a digital facility such as the one Google is developing. The scope of the problem is frequently compared to the destruction of the ancient Library of Alexandria. ICTs enable a possible solution, but the questions of ownership and copyright complicate a resolution.

Another example of the current difficulties of defining ownership occurs in the popular music industry. A hundred years ago, popular music was copyrighted and distributed exclusively on printed sheets. People who pirated the music and sold it for their personal profit could be sued and fined. Some offenders were even jailed. As the production technology of capturing the sounds of performances advanced from analog impressions on wax cylinders to vinyl disks, a new industry of recording, distribution, and sales emerged. As that technology progressed further from records to tapes and CDs, the recording industry became an annual multi-billion-dollar enterprise, and fans strived to acquire the latest popular music albums. Performers became public celebrities, and the ancillary support industries of advertising, marketing, public relations, and concert tour management flourished.

Once the Internet reached the stage of allowing high-speed transmission and relatively easy downloading capabilities, the dynamics of the music industry shifted quickly and dramatically from the acquisition of albums to the accessing of copies of single songs. Sales of albums plummeted, and illegal downloading and sharing of songs skyrocketed. The once-substantial revenue stream of all the participants in the recording industry from the performers to the marketers rapidly declined. From 1999 to 2001, Napster, a company created by college student Shawn Fanning, offered a free exchange of music files. It achieved immediate success with 25 million users and more than 80 million songs. However, the company was shut down by a court order for massive copyright violations. With leadership from the Recording Industry Association of America (RIAA), the industry swung into action and filed lawsuits against those individuals it could identify as the most egregious illegal copyright offenders. Not surprisingly, most of the offenders were high school and college students who downloaded songs and freely circulated them to their friends. Apple now offers iTunes, which allows the purchase and download of single songs for a nominal fee, and its success seems to have stemmed the tide of massive illegal downloading and sharing. However, the financial health of the music industry has not returned, and its future is by no means clear.

The motion picture industry has undergone a similar transformation. With the proliferation of ICTs, the industry that arose during the twentieth century to create, distribute, and project motion pictures in theaters is facing similar problems. When the speed of Internet file transfers reached sufficient capacity to download digital versions of feature-length movies in just a few minutes, the illegal copying and distribution of films also became widespread. As with the music industry, the Motion Picture Association of America (MPAA) began filing suits against the perpetrators. Inspired by the writings of such scholars as Lawrence Lessig, a law professor at Harvard University, a movement emerged with strong support from students promoting what they called *free culture*. They espoused that all copyright laws and practices needed to be revised to promote the free distribution of all intellectual products, including music, films, and all forms of artworks. The costs, rewards, and incentives to support all the production and distribution would come from online advertising on the download sites. Lessig claims that the existing copyright system makes most young people scofflaws for they regularly violate the law. As one might expect, neither the RIAA nor the MPAA support the free culture proposal.

Professor John Tehranian of the University of Utah Law School uses a strict interpretation of copyright law to estimate that his unpaid royalties for one year amounts to over $4.5 billion. This includes his and his students' use of email and Internet websites plus the copyright-protected image of the tattoo on his shoulder. The tattoo is displayed publicly every day when he uses the university swimming pool for exercise. Tehranian points out that either he is a model scofflaw for his students, or copyright law does not mean what it says.

As mentioned earlier, the fact that so many varied artistic productions are recorded in digital format readily enables creative works or mashups that combine different types of media representations. Multi-media productions can be created with nothing more than access to a personal computer and the Internet. Consequently, ICTs facilitate a huge increase in the production of artistic and creative works and make it possible to distribute these works at practically zero cost. For the first time in recorded history, it is possible for creative artists to reach huge audiences and to combine elements from previously produced artworks, including music, films, videos, paintings, sculptures, and text to produce new, media hybrid, intellectual works.

How does one acknowledge the copyrights of works that are partially excerpted and combined with other elements? The idea that this should not be necessary and that intellectual products of any genre should be made available cost free to all members of a society is not new. Many individuals and groups have proposed such a concept throughout the history of civilization. Some utopian communities, for example, promoted increased cultural production and consumption as a generally recognized social objective. Clearly, there is a need to revisit the basic assumptions underlying the protection of the rights of artists and at the same time to recognize and support the new capabilities for expanded creativity and cultural promotion.

The Open Source Movement

Finally, under the rubric of ownership the concept of *open source* as facilitated by ICTs has become salient, particularly with reference to computer software and patents. As with free culture, open source is not a new concept. However, ICTs, and particularly the Internet, greatly facilitate the open source movement, and the concept now takes on additional import and impact. The concept of ownership that emerged during the Industrial Revolution kept the knowledge of production methods as carefully guarded secrets. The recipes for producing soft drinks by bottling companies, the procedures for manufacturing drugs by pharmaceutical companies, and the techniques for making computer chips have all been carefully protected by industrial secrecy. The zeal for such secrecy spawned a small ancillary industry called *industrial espionage,* in which hired agents tried to break into manufacturing sites to steal the requisite know-how.

The open source movement received substantial impetus from the creation and spread of the computer operating system Linux. Developed by Linus Torvalds, a Finnish software engineer in the early 1990s, Linux is a free, open-source operating system that can be used by a wide variety of computing devices, ranging from desktop computers and smart phones to wristwatches. However, its distinguishing feature is that it is available for use and modification by everyone at no cost. For example, the user of virtually any computing device can download at no cost a system to operate any available software programs or application. This kind of open source system is in direct competition with commercial software producers, most particularly Microsoft and Apple. These corporations and many others charge hefty fees to obtain and modify their operating systems. Furthermore, the user can access and modify any part of the Linux operating system. It takes a substantial level of technical expertise to make such modifications, but the system is available to the public at no cost.

A number of companies now recognize the wisdom of offering industrial secrets to everyone; they believe that by making all their activities transparent to the public, they make it possible for others to develop new products that will be complementary to theirs, thereby creating collaborators. In the information age, more people recognize that collaboration can be a powerful vehicle for growth. Sometimes that collaboration will entail a buyout. Apple has experienced phenomenal expansion and financial reward by collaborating with, and frequently buying, other companies to develop applications for its many varieties of computers and handheld devices. IBM recently made some of its patents open source and available at no cost to the public, and it expects that this will be a more prudent corporate policy for future growth and development. Growing evidence suggests that in the corporate world as well as in the artistic world, we need to revise our concepts of ownership and the management of digital copyrights within the framework of normative order.

PRIVACY

The concept of *privacy* has taken on new meaning and salience as a result of the proliferation of ICTs. The *Oxford English Dictionary* defines *privacy* as "The state of being withdrawn from the society of others or from public attention; freedom from disturbance or intrusion; seclusion." The Fourth Amendment to the U.S. Constitution guarantees that the privacy of its citizens is protected from "unreasonable search and seizure." Although concepts of privacy vary substantially from one nation and culture to another, the common theme is that individuals have the right to keep information about themselves from public scrutiny. However, effectively limiting the collection, distribution, and sale of personal information is difficult, and shortly it may become impossible. From credit data banks and online searches and purchases, detailed profiles can be created on most people. Add to those sources information from court proceedings, background investigations, and educational or military records and privacy could be under serious attack. A resolution of this problem will require collaboration among government agencies, the private sector, and citizen advocacy groups.

As mentioned in Chapter 3, the recent disclosure that the National Security Agency (NSA), a U.S. intelligence agency, has been accessing the data files of companies that record the telephone calls and Internet activities of all their customers has caused an uproar from those who wish to defend their privacy right as defined by the Fourth Amendment. The intent of the NSA is to recognize patterns of communication that might occur among terrorists to prevent their attacks. The resolution of this conflict between those who claim that governments must develop new methods to combat terrorism and those who promote the need to protect privacy will not be resolved easily.

Uses and Misuses of Large Databases

The creation and maintenance of large databases of information about individuals became possible only when computers with large storage capacity were developed. One of the first to propose such large databases was the late Paul Baran, an engineer at the RAND Corporation in Santa Monica, California in the early 1960s. His concept of an information utility as a resource for government and business policy formation was widely discussed, and it generated concerns over potential privacy violations. Incidentally, Baran was one of the developers of packet switching, the technology that enabled the ARPANET and eventually the Internet.

Almost all businesses and government agencies now maintain large databases, and unfortunately many have become easy targets for hackers. These databases are stored either locally on sites with large memory servers or increasingly in remote servers that are accessed through the Internet in what is referred to as cloud computing. Both methods are subject to hacking, and they frequently lure unscrupulous persons to steal personal records and sell them to others for illegal purposes such as identity theft. Almost daily reports document the frequency of stolen data from such sources as retailers, banks, motor vehicle agencies, and universities. Estimates of identity theft substantially increase yearly, and many experts attribute this pattern to the hacking of large databases. One of the arguments against health care reform in the U.S. is that a nationalized system would create individual electronic medical histories that would be difficult to protect from hackers. Indeed, opponents of universal and centrally organized health delivery cite potential privacy invasions as a sufficient basis to continue with the current privatized system and its ever-escalating costs.

The tracking of online consumer behaviors is another and more recent attack on privacy that is enabled by ICTs. Most online purchases result in a permanent record. When combined with other purchase records, a profile can be created and used to target advertising to generate additional sales. Advertisers find this a lucrative practice, and they actively pursue buying these records from a variety of online merchants. They then combine the files using credit card numbers and shipping addresses to create an informative and thus valuable customer profile. Furthermore, some vendors' websites record the path of searches of their merchandise. An online bookseller might keep records of both books purchased as well as the books that a potential customer reviewed but did not buy. In this manner, the vendor can create a rich profile of reading habits and interests. All this tracking is driven by the hope of selling more and more products online.

Social Networking

A second form of tracking occurs with the patterns of usage and the content of social networking systems. ICTs enable a wide variety of these systems, and they are enjoying enormous growth. Connections to these networks can be readily obtained through the Internet using a wide variety of devices, including personal computers, notebooks, cell phones, and many handheld units. Networks such as Facebook, MySpace, and Twitter report registered users in the hundreds of millions. Facebook alone accounts for more than 500 million subscribers. Popularity among young people, especially preteens and adolescents, is rampant, and the exchange of brief text messages and photographs is growing rapidly in part due to the minimal costs. As mentioned earlier, the users are developing a unique vocabulary consisting of letters, numbers, and acronyms to convey their messages concisely. The dictionaries of these texting languages are becoming quite extensive. Many young people exchange messages and photographs of themselves or others that are often misread as child pornography, and they seem to be naïve in underestimating how quickly and widely such messages can spread on the Internet with negative consequences. Some employers routinely search Facebook for files of people they are considering as future employees. Local school authorities and law enforcement officers have taken disciplinary actions against the originators of such risqué texts and pictures. Other reports indicate that students are using these networks for online bullying. Several such recent incidents have resulted in the attacked people taking their own lives out of desperation.

Not all social networks create such problems. Many professional and adult networks are proving both useful and popular. For example, LinkedIn, a professional network that is used to connect with others in the business, finance, education, and management communities, claims to have more than 225 million users registered in more than 200 countries and territories as of 2013. Countless other networks are organized around various professions, occupations, interests, and backgrounds. Colleges, universities, and professional schools are creating networks to facilitate alumni connections and to fundraise. As mentioned earlier in the previous chapter in the discussion of political processes, a large number of social networks are organized around social and political causes. The networks created during Barack Obama's campaigns for the U.S. presidency in 2008 and 2012 were instrumental in recruiting staff and voters who helped achieve his victory. However, each of these networks has the potential of revealing information about its members that can infringe on their personal privacy. The risk becomes substantial when the data from multiple sources are combined.

Protection

Prior to the advent of ICTs, privacy was not a concern for most people. In past years, famous cases of egregious violations of individual privacy did not promote the kind of general anxieties now evident about the growing data files tracking patterns of consumption, financial records, court proceedings, and more. These files are readily available to those with the requisite technical expertise to hack the databases. Evidence of this growing public anxiety can be found in the rapid emergence of the new profession of privacy protection. Professional legal specialties now focus exclusively on privacy management. They offer advice to corporations, organizations, families, and individuals on how they can best protect themselves from privacy invasions, character defamations, and identity theft.

Radio frequency identification (RFID) is another type of ICT use that bears discussion in the context of privacy. Originally developed as a means of inventory control in manufacturing and distribution functions, RFID uses small chips that are attached to various products to keep track of their locations as they move from manufacturing to shipment to inventory storage and on to final sales or disposition locations. Chips vary in their capacity to both send and receive radio signals and in the distances the signals can travel. If the chips are attached to consumer products and these records are combined with data from other sources, a rather complete profile of customers' purchases and background information can readily be compiled. Furthermore, the Food and Drug Administration (FDA) has approved implanting RFID tags in animals for inventory control and in humans for biomedical purposes. Consumer protection groups are worried about the use of RFIDs, particularly when people are not fully informed of their presence. They are especially outraged at the prospect of widespread human implanting. ICTs and RFIDs are creating the need for further reflection on the appropriate use of technologies. The need to protect citizens from unscrupulous behaviors and exploitation by both government and private business interests is widely accepted.

An interesting research project at Princeton University raises important policy issues relevant to privacy protection. Karen Levy, a doctoral candidate in sociology, has spent several years interviewing people involved in the trucking industry, including drivers, supervisors, and company executives. At the present time drivers fill out a handwritten form reporting all their trips with miles and times. Drivers and supervisors recognize that this is an honor system relying upon accurate reporting, and yet everyone accepts that there is some leeway acceptable in precise reporting. A new system of recording the data is under development by a commercial technology company. Using a global positioning system (GPS) the company has developed the Electronic On Board Recorder (EOBR) to document continuous data about truck locations and movements. These data are then transmitted to a central office to create permanent records. The U.S. Department of Transportation is considering a regulation that would mandate that all interstate truckers install and use EOBRs.

Such a system obviously would be appreciated by many trucking company managers. Equally obvious are the negative reactions reported in interviews conducted by Levy of a sample of drivers. One of the current attractions of this occupation is the relative degree of freedom afforded the drivers without the kind of direct supervision that exists in the majority of workplaces. Levy reports that many older drivers interviewed viewed EORBs as an invasion of their privacy, and they threatened to resign their positions rather than submit to this kind of electronic surveillance. On the other hand, many younger drivers were angry that such a system seemed inevitable, but they indicated that they would simply have to adjust their expectations as to the nature of the workplace.

This project is interesting, because of the implication that many other occupations and professions may soon adopt similar systems that employ ICTs to record how people perform in their work roles. For example, such a system has been installed in hospital and medical offices to monitor the activities and efficiency of doctors and nurses. Furthermore, if such recording systems were to be coupled with surveillance cameras, the invasion of privacy in the workplace could become widespread. It seems reasonable to expect that such animosity among many occupations and profession would produce a growing sense of alienation. As mentioned in Chapter 2, sociologists call this *anomie;* and it does not promote social cohesion.

RELIGION

Religious activities and organizations have always played an important role in the creation and maintenance of normative order in all societies. In most instances, the tenets of religious faith reflect the same values that provide social order. Therefore, it should come as no surprise that religion as a major social institution will also undergo transitions as a consequence of the proliferation of ICTs. Perhaps the most obvious is the portrayal of and participation in religious services using media afforded by the Internet, instead of physical church attendance. This type of remote participation in services is known as *distance religion*. Online church services can be viewed on a variety of output devices, including personal computers, notebooks, handheld devices, and even cell phones. Many churches are making it possible for their distant congregants to make contributions to regular church collections by using a credit card over the Internet. Such broadcasts allow people to attend church when they might otherwise be unable to do so because of physical limitations or excessive distances from the place of the worship service. Many services are recorded and played back at another time or even repeated. This concept of asynchronous church services appeals to many who prefer not to attend a service at an appointed time.

Religious services have a long history with radio and television broadcasting. The activities and financial success of the televangelists are well-known. However, religious services and practices now enabled by ICTs can generate heightened aesthetic experiences involving viewing, listening, and now participating. Some online religious activities involve virtual visits to sites that are considered as especially holy in various faiths, such as the Western or Wailing Wall in the Old City of Jerusalem. These activities can also encourage praying, meditation, and reflecting on religious icons. Reports of such aesthetic experiences in online religious activities come from all the major religions of the world, including Christianity, Judaism, Islam, Buddhism, and Hinduism.

Virtual religious communities have begun to appear. A recent book is *Digital Jesus: The Making of a New Christian Fundamentalist Internet Community on the Internet*, by Professor Robert Glenn Howard, University of Wisconsin–Madison. It reports an ethnographic study of two online religious groups that are directed by persons who have an interest in promoting particular interpretations of the Bible and exploring the relationships between God and humans. Both groups have existed for some decades and began with email lists, but the current capabilities of ICTs have expanded their memberships and activities into an ongoing online religious community.

The virtual community Second Life has many avatars that are active participants in a variety of religious services. In the book entitled *Tweet If You Love Jesus: Practicing Church in the Digital Reformation*, Dr. Elizabeth Drescher, a faculty member at Santa Clara University, proposes that ICTs are ushering in a new reformation of the Christian churches. This reformation enables additional forms of communication for religious activities that can appeal to regular users of ICTs. Drescher draws comparisons between the impact of the printing and distribution of sacred texts on the Protestant Reformation in the early sixteenth century and the proliferation of ICTs on current developments in religious activities and organizations. Interestingly in the title of Drescher's book, an icon of a heart replaces the word "love," a clear appeal to those who use icons in their online communications.

One of the major functions of most religions is to provide support to members during major life transitions such as births, coming of age rituals, marriages, and deaths. Increasingly, people can participate in such events from a distance using the Internet to support visual and audio signal transmissions. People are attending baptisms and weddings of friends and relatives using *telepresence*. This practice is becoming common in the rituals that accompany death as well. Viewings, wakes, memorial services, and even burials can be video recorded in real time and transmitted over the Internet to allow virtual attendance by people at a great distance. In addition, people find it comforting to create online equivalents of tombstones that contain photographs or videos and important facts about the deceased. Frequently, these virtual markers will contain several extended eulogies. Such practices are facilitated by funeral directors, and they can provide solace to those who are grieving.

A recent report details how a large number of friends of a young woman who was killed in an automobile accident managed their grief. They created a Web page in her memory and regularly posted messages to other group members, frequently addressing their comments to their deceased friend. Posting to this site continued for several years, and the friends found it an effective grief management experience. In the future, perhaps support for other emotional needs may be similarly facilitated by ICTs.

Finally, church officials are now using ICTs to communicate with their congregants. The bishop of a large, urban diocese of the Roman Catholic Church launched a weekly blog to communicate with all his parishioners. This is the equivalent of the traditional printed newsletter with a major exception. The bishop invites his parishioners to respond to his blogs, and exchanges frequently occur between the church official and his flock. In this instance, ICTs have enabled a new kind of relationship and dialog between the clergy and laity. In addition, some clergy use the Internet and email instead of making house calls to members of their congregations. One of the traditional tasks of the clergy is to visit the homebound or the elderly, and the ICT-enabled house call can make it more efficient for the increasing numbers of senior citizens who are becoming computer literate and desire the attention of their religious leaders.

CRIMINAL BEHAVIOR

As mentioned in the introduction to normative order in Chapter 3, deviant and criminal behaviors have always existed in all societies. Many perpetrators have not been adequately socialized to the prevailing norms; others are promoting different values or perspectives that may be harbingers of the future, as has occurred in the arts. What may be considered a shocking exhibition of artworks when first displayed may, in the not too distant future, be recognized as path breaking and innovative.

The widespread use of ICTs enables both more efficient modes and new forms of deviance and crime. There is a difference between using ICTs to commit crimes, for example, stealing another person's identity and withdrawing funds from a bank account, and making an ICT the object of a malicious attack, such as destroying operating systems or data files. Identity theft is the foremost and fastest growing ICT related crime. Assuming the identity of another person, consuming his or her assets, and creating new liabilities have become much easier to achieve with ICTs. Accessing the requisite information such as credit card data, Social Security numbers, and addresses and then using this information to make fraudulent charges or to open new lines of credit are relatively simple tasks over the Internet. There are now services that will scan the Internet for evidence of theft and notify the subscriber for a monthly fee. Some services also offer insurance against losses attributable to a failure of their scanning services.

An interesting case recently occurred at a U.S. public university in the Midwest. Two students were charged with a vast identity theft scheme. Using information gathered by hacking into the files of online retailers, auction sites, and commercial banks, they had opened several hundred false identity accounts. They then purchased items and arranged to ship them to their home for sale at greatly reduced prices. Although the activity was relatively small on any one false identity, by using several hundred such accounts, they accumulated several million dollars in profits before being apprehended and shut down. Such crimes contribute substantially to widespread fears of the lack of security in online commerce.

Gambling is legal in most countries if practiced in specially designated and licensed casinos. Government-operated lotteries are also quite common and, of course, legal. However, additional illegal gambling operations have always existed from neighborhood numbers games to betting on sports events. The numbers games still operate on a small scale, but the government operated lotteries have taken up many of their former clients. In addition, betting on sports events has expanded its scope by using ICTs for both the placing of bets and the paying of winnings. The Internet also enables a new kind of gambling that appears to be popular among college students and young adults. College students have reportedly accumulated huge winnings and losses by playing online poker. Exactly where the headquarters of these online card parlors are located is unclear, but they seem to be in smaller countries that have either little or no casino oversight and tax regulations.

Tax evasion is another form of illegal behavior enabled by ICTs. In the offline world, one is obligated to report as taxable income any monies received from sales in excess of the cost of the item plus expenses. Many jurisdictions also require that sales taxes be paid. However, sales and purchases of items on Internet sites are completely unregulated and not readily viewed by tax agencies. No one knows how much sales tax revenue is lost to local and state governments in this largely covert economy, but it must be enormous. Legislation intended to make sales tax payment mandatory has been proposed, but there appears to be limited support.

ICTs enable greater production and distribution of pornography. It is estimated that more than 50 percent of all Internet traffic at any given moment contains pornographic material. Legislatures and courts in the U.S. have left the nature of pornography vaguely defined. Various rulings in different jurisdictions have resulted in great ambiguity. However, as mentioned in Chapter 3, there is near universal agreement that pornography that employs children is immoral and should be illegal. Law enforcement agencies vigorously pursue any leads that might uncover producers, distributors, or consumers of child pornography. Even possession is a crime punishable by imprisonment in most jurisdictions.

Crimes that involve illegal activities across national borders have also benefited from the international spread of ICTs. Communications among national branches of organized crime groups are enhanced using the Internet, and the exchange and laundering of illicit funds are facilitated. Unfortunately, it seems that communications and collaborations among national and international law enforcement agencies lag behind the activities of organized crime. ICTs also facilitate the illegal activities of terrorist organizations. Reports indicate that extremist Islamic groups are quite sophisticated in their use of the Internet and email to coordinate their covert activities. Clearly, international cooperation on a scale that has not previously been possible will be necessary to contain both international organized crime and terrorist activities, and such collaboration is not likely to happen in the near future.

ICTs also enable plagiarism, a criminal act. Plagiarism involves creating unauthorized copies of materials and distributing them without permission of the copyright holder. Frequently, the copying involves selling the materials at a much lower price than in the retail marketplace. Publishing and media companies in the U.S. charge that this practice is widespread in China, and they have had little success either in halting it or in being reimbursed.

A variation of plagiarism is increasing among high school and college students. Although it may not be illegal, it is at least dishonest. Students purchase materials from commercial services and use them without attribution to complete written assignments in their courses. Again, this is not a completely new practice. For years, student groups saved copies of old papers and tests to assist their younger peers. However, the Internet has made it possible to provide essays to students around the globe on a more extensive basis than previously imagined. One service boasts it has more than 1 million complete essays in its files for sale on every subject covered in a typical undergraduate curriculum. Interestingly, several services now offer a way for professors to compare students' current essays to their extensive files to see if plagiarism has occurred. Although the students' papers need not be registered for copyright protection, a violation could occur here. The essays sent by the professor may violate the original students' copyrights, for the commercial company keeps all essays submitted and adds them to its ever-expanding files. The students receive no remuneration. Are the professors and the companies jointly violating the students' rights of ownership of their essays? It appears that the companies receive the papers free, yet they charge for the searches for possible plagiarism.

As mentioned earlier, the second means of using ICTs for criminal activities includes attacking the computers and networks for the purpose of corrupting and destroying the operating systems and data files. In recent years, reports of such attacks have increased, some confirmed and others alleged, but not proven. For example, the U.S. military services have claimed that unknown parties hacked into the worldwide Pentagon computer system, corrupting both programs and data. Officials suspect that the attacks originated in China or with terrorists groups, but there have been no reports confirming the suspicions. Several Balkan countries have also reported such attacks, claiming they have evidence that implicates Russia. Of course, it is one thing to identify the location of the attacks, but that does not determine whether the perpetrator is affiliated with any group or government or simply a private hacker.

Espionage can also be facilitated by breaking into ICT systems. No one knows how extensive such activities are, but it seems reasonable to assume they occur and probably are increasing. The recent cases of WikiLeaks and the violations of privacy attributed to journalists employed in the Murdoch news organization are clear examples of attacks on the security of ICTs. Perhaps most troubling is the large number of people implicated in such activities. Recall that in the past, huge public denunciations have protested the creations of many artists. In just a few short years, many of those same creations were hailed as important innovations. Will the hacking of ICTS become so widespread and commonplace that it will be widely accepted as ethical behavior?

Both Chapters 3 and 4 examined how ICTs are influencing changes in normative order, ranging from the division of labor and political processes in Chapter 3 to ownership, privacy, religious practices, and criminal behavior in Chapter 4. The following chapters use a similar format for exploring how ICTs are changing other major social institutions, including socialization and education, the creation and utilization of knowledge, the expansion of the consumer economy, and the altered modes in which time and space structure our daily lives.

5 Learning

Among all the major social institutions, perhaps the most dramatic changes afforded by information communication technologies (ICTs) are those that socialize and educate children. Not too many generations ago, children were prepared for their future adult roles solely by the members of their extended families in multi-generational farming communities. In many countries, the family and the church shared the responsibility for inculcating the prevailing values. As public elementary and secondary education emerged, many children received additional civic and moral education from school personnel. Nevertheless, until a century ago, the family was still the major force in socialization, and the process consisted primarily of observing and emulating the exemplary behavior of the older relatives.

Since the Industrial Revolution, which first emerged in the late eighteenth century, these patterns of socialization have been gradually modified. By the midpoint of the twentieth century, the adolescent peer culture began to exert substantial influence on the values often in conflict with those of the adults. With the advent of ICTs in the past 50 years, the rate of change is escalating; now in the new millennium, the Internet has become a primary agent of socialization for many children. The values traditionally promoted by the family, the church, and the schools are being replaced by a subculture transmitted via ICTs with little adult knowledge of or control over the content. Rather, commercial interests nurture the adolescent peer culture, which is the primary determinant of the values of this new mode of socialization of the young.

This chapter explores the changes that have occurred and are ongoing in the processes of socialization and education. The chapter consists of three sections: the socialization of young children and adolescents, the teaching and learning in elementary and secondary schools, and the structure and functioning of higher education. All the innovations and changes discussed have important but not always encouraging implications for the finances of the institutions that prepare the future adult citizens of the world.

AGENTS AND CONSEQUENCES OF SOCIALIZATION

The replacement of the family as the major socialization agent of children began as families started moving from farms to communities near the factories where they were increasingly employed. As industrialization progressed and more people left the farms, the pattern of the extended or multi-generational family gave way to the nuclear family, consisting only of parents and their children under the same roof. The children then had fewer adults living with them to serve as role models. Eventually, both parents worked outside the home, and children spent most of their day in school interacting with their teachers and fellow students. Supervision was provided by teachers and other school staff, and the influence of the family on the children gradually diminished. These transitions took place over several hundred years and multiple generations, and the changes in socialization agents from family to school and peers were more commonly found in the gradually expanding urban and suburban areas near the centers of industrial production.

Peer Cultures and Internet Content

By the middle of the twentieth century, peer cultures among children and adolescents had become a prominent feature of growing up in most North American and European countries. By the end of that century, the pervasive influence of peer cultures had spread to most of the developed nations around the globe. The social networking enabled by ICTs became a major force in the expansion and influence of those peer cultures. Furthermore, the social networks are almost exclusively devoid of any adult input, participation, supervision, or even awareness. The socialization process was once the exclusive province of a limited number of adults, including family, teachers, and clergy. However, socialization now seems to be very much in the exclusive control of the younger age cohorts, the managers of social networks, and the commercial interests that provide advertising revenues.

It seems clear that the vast majority of middle and high school students use mobile, handheld devices to connect to social networks to sustain frequent and close contact with their circles of friends. A random survey of adolescents conducted by the Internet & American Life Project of the Pew Research Center revealed that in 2012, 95 percent of teens between the ages of 12 and 17 had access to the Internet. According to an earlier Pew survey, social networking sites were the most commonly used Internet capabilities among all adolescents. It now was clear that all the social networking software had become the new, powerful, and pervasive agents of socialization for many children.

A further complication of the absence of adult involvement in or supervision of children's exposure to the content of the Internet and social networks is the corresponding increased involvement of parents in the 24/7 global economy. Harriet B. Presser, professor of sociology at New York University, has competently described the many difficulties of contemporary parenting. Many parents are continuously connected to various networks for online work or social interactions, and this pattern infringes on their time to be involved with children. Consequently, because adults are so involved in sending and receiving messages or surfing the Internet, their availability to serve as agents of socialization for their children is severely compromised. There are reports that children feel that the significant adults in their lives are simply not available. Both children and adults are constantly online.

In an article published in 2001, Marc Prensky called the current teen and young adult cohorts "digital natives," for they grew up in a time when ICTs were already ubiquitous. These young people are comfortable surfing the Internet and using all kinds of technology, including computers, DVDs, cell phones, and iPads. Rather than formally learning how to use these technologies, they seem to absorb them as readily as they learned to speak. Prensky contrasts the digital natives with the "digital immigrants," the older cohorts who had to learn to use all these technologies that did not exist when they were children. As with geographical migrants, the digital immigrants are struggling to learn the vernacular and mores of cyber culture. While the digital immigrants are busy adapting to the new culture, the natives are wandering in cyberspace exposed to a wide variety of materials, some of which are of questionable value.

There is a paucity of valid and reliable data describing the content of the Internet. This is, of course, understandable, because websites are constantly being created, deleted, and modified. To keep current on the numbers, let alone the content, is daunting, if not impossible. However, it is possible to describe the kinds of content that are widely distributed and question their value to the development of young people. For example, a large amount of pornographic material now exists on the Internet. As mentioned earlier, one estimate claims that at any given time, 50 percent of all Internet traffic is pornographic. The creation, distribution, and viewing of pornography are not new. However, widespread distribution of pornography is new, and it is the Internet that makes it possible. All children can now easily access and repeatedly view the most hardcore materials, and we should assume that many do.

How injurious are such experiences to the health and development of children? Does viewing pornographic materials eventually lead to acting out and sexual violence? Convicted sex offenders have reported that viewing such materials led them to commit violent sexual acts. However, this is a question that cries out for empirical research to guide policy formation regarding the accessibility of pornography. Again, this is not a new issue; the courts have long struggled to find operational definitions of pornography. Protection of the public, especially children, is a legitimate concern that is counterpoised to the right of freedom of expression. The intensity of this dilemma increases when the persons portrayed in the pornographic materials are themselves children. The courts and law enforcement agencies in the U.S. are vigilant in their pursuit and punishment of persons who create, distribute, or even possess pornographic materials involving children.

A growing concern is that many young people use the Internet to engage in behaviors that put them at risk of being harmed or exploited. Numerous accounts exist of young people participating in discussions in online chat rooms where the conversations have sexual content. Subsequently, they may meet off line in face-to-face encounters and discover that their newly found friends expect sexual liaisons. Furthermore, these contacts frequently turn out to be adults, and the child can be at risk of serious harm. Risky behaviors among adolescents and young adults certainly are not new. History and literature are replete with accounts of the foolish acts of youth. The ancients Greeks complained about the reckless behavior of their youth and promoted plans for containing their perceived excesses. However, today ICTs provide a new and far-reaching vehicle for risky behaviors that can be observed and participated in by a very large number of the young as well as adults.

For example, the Pew surveys mentioned earlier showed that many young people regularly and incessantly exchange messages with friends using mobile, handheld devices. This exchange is known as Instant Messaging (IM) or texting. Most new handheld devices include digital cameras, and the users frequently send "selfies," photographs or videos of themselves. When young people send sexually provocative pictures of themselves, called *sexting,* they run serious risks. All too often the pictures get widely circulated and can exist for a very long time, potentially causing embarrassing consequences in the future. Such pictures may legally be defined as pornography. If the subject of the picture is younger than age 18, anyone who creates, distributes, or views the picture is committing a felony crime that could result in a prison sentence. As mentioned earlier, risky behavior of the young has always had the potential for long-lasting negative consequences. However, sexting with ICTs brings that behavior to the attention of a vastly larger audience, thereby increasing the probability of negative sanctions. Of relevance here is the vast public attention paid to the sexting activities of a former member of Congress who attempted to run for the office of Mayor of New York City.

Closely related to the prevalence of pornography is the amount of violence portrayed on many websites to which children also have easy access. Years of research have established a clear connection between children viewing violence portrayed in media and subsequent use of physical force against others. The film and TV industries have rating systems that alert viewers to violent and sexual content and provide guidelines as to appropriate ages for viewing. The rating systems have been criticized as inadequate and misleading. However, they do exist. On the Internet, some websites may require a credit card charge, and others simply ask the users to state that they are older than age 18. Such barriers are easy to overcome, and children quickly learn and share with their peers the techniques to access the most violent sites.

Massively multi-player online role-playing games (MMORPGs) allow hundreds of thousands of players around the globe to participate simultaneously. Many of these online games portray an array of violent behaviors, and they frequently involve the virtual shooting and killing of many presumed enemies. Armed with an arsenal of virtual weapons, game players accumulate a score for the number of enemies slain before they are killed themselves. Some MMORPGs are free, and others charge a fee. A popular and financially successful online game, Grand Auto Theft (GTA), simulates automobile thefts and is loaded with violent and sexual scenarios. It is a spinoff of a 1977 film, and it now has many different versions. Repeated participation in such virtual violence concerns many parents and educators, who worry that those children may become immune to actual violence and suffering. As with pornography, concern is magnified when the actual number of children who play such games is unknown. However, the potential impact of the values displayed in the violent and sexual behaviors on the socialization of the children is of great concern.

Bullying, Commercial Advertising, and Cognitive Deficits

Bullying is another form of violence among children that is enabled by ICTs, and it has a potential negative influence on socialization. Bullying among children is a common behavior that has probably been going on since time immemorial. In most instances, the peer culture frowns upon reporting bullying incidents to any adults. The prevailing norm among the children is that the involved parties should work out their differences without adult intervention. In recent years, many teachers and parents have reported increases in bullying behaviors using email and the Internet. It may be that online bullying is more devastating for the victim, because copies of emails and addresses of websites can be circulated to a much larger number of peers than would otherwise witness a public bullying encounter. The spectacle of being bullied makes the event ever more embarrassing. As noted in Chapter 4, several children have committed suicide after especially vicious bullying incidents.

Online bullying is especially dangerous, because it takes place where adults are most unlikely to observe it. At least schoolyard bullying occurs where some school staff may see and report it, but online bullying is likely to be observed only by other children. The more pervasive negative impact of online bullying affects the children who see the offensive behavior go unrecognized and unpunished. Although bullying is not new, a much larger number of children watch it happen with little or no consequence for the bullies. When children observe behavior that violates the prevailing normative order and there are no negative sanctions, effective socialization is threatened.

In yet another dimension of online activities, young people are pervasively and incessantly exposed to commercial advertising. Most of the activities that pre-teens and teenagers engage in online are free. However, the hidden price for these activities is the advertising, which is distributed over most websites and accompanies most social interchanges. Advertising revenue supports most of the economy of cyberspace. The producers of goods and services believe that by advertising online, they will increase sales and revenue, and they are willing to pay substantial amounts for premium positions on websites. Of course, the same economic structure supports the TV industry, and the prevalence of those advertisements is a commonly accepted aspect of today's culture. However, if one adds the number of hours that children are exposed to such messages on TV to the additional advertising encountered online, the socialization of children now includes a powerful message to engage in extensive consumption.

It is appropriate to ask whether the socialization of children should include encouragement of consumption as a major value of normative order. Some would argue that ever-expanding consumption is a cornerstone of a modern capitalist economy, and, therefore, socialization should encourage the development of future consumer behavior. Others would argue that unchecked consumption is sustainable neither for the economy nor the environment. This discussion does not promote either position. However, it does point out that the online activities of children expose them to consumer-oriented socialization. Furthermore, this development takes place with little public awareness and debate as to the utility and adequacy of such socialization. It has already saturated the print and TV media, but the rapid and pervasive adoption of online activities should increase public awareness of this issue.

A final observation concerning a potential negative impact on socialization is that ICTs promote short attention span activities. Social networks encourage brief communications. Twitter limits exchanges to 140 characters. Many users of ICTs report that they frequently multitask, simultaneously engaging the following activities: surfing the Web, sending emails, completing homework, talking on the telephone, sending text messages, watching TV, or listening to downloaded music. Some cognitive scientists doubt that true multitasking is even possible. Rather, they argue that conscious attention can be focused on only one object or process at a time. And they claim that what is referred to as multitasking is actually a rapid switching of short bursts of attention among two or more activities. Of course, some activities may be accomplished with minimal consciousness, but the point here is that true parallel cognitive processing is not humanly possible. Some people, especially younger ones, may give the appearance of multitasking, but it is only an illusion. However, the pattern of multiple short attention focusing may become dominant and result in limiting the capacity for sustained cognitive focusing. Recall that the ancient Greeks warned correctly that reading and writing would limit human capacity for memory.

In 2008, Nicholas Carr, a prolific writer and astute observer of practices in business management, technology implementations, and cultural phenomena, published a provocative article in *The Atlantic*, "Is Google Making Us Stupid?" He posited that the cognitive changes generated by multitasking on the Internet, social networking, and more pointedly Google were making us all stupid. The article generated many intense reactions, particularly from the computer hardware and software industries. It obviously raised a sensitive question, one with which many people could identify. Short attention span is a cognitive deficit, and it can preclude extended deliberation and meditation. As the use of the Internet and social networks is so widespread and pervasive among children, attention deficit disorders reinforced by ICTs could become a major problem. Many young people report that they rarely read anything off line, and they describe their online reading as skimming rather than careful examination of the contents of the text. With the proliferation of touch screen and handheld devices, young people say they are more comfortable communicating with icons than text. A first-year student at a highly selective college recently reported that he had never in his life read an entire book. He lost interest after a few pages and returned to texting his friends. All these reports and patterns of communicative behaviors portend substantial changes in the ways we communicate with one another. Furthermore, they will alter the patterns of future socialization and communication.

Potential positive impacts of ICTs on the socialization of children also need to be acknowledged. For example, ICTs can support tutorials on topics dealing with various aspects of sexuality that parents and other adults have difficulty communicating to children and adolescents. In North Carolina, an online service answers questions submitted anonymously. A staff of professional counselors is on call at all times to provide advice for young people who submit requests for information and advice. To avoid conflict with people or organizations with opposing views, staff members carefully avoid taking positions on such issues as contraception and abortion. The central theme of their responses is to reassure the callers that their questions and concerns are legitimate and appropriate. They provide reassurance and typically guide the person to seek further local help. The anonymity of the online system allows young people to overcome any embarrassment they might feel if they attempted to have such a conversation in a face-to-face setting. In some respects, this kind of anonymous, ICT-enabled interaction resembles the dynamics of an individual psychotherapy session.

TEACHING AND LEARNING IN SCHOOLS

Immediately after the first large-scale digital computers were introduced in the 1960s, many people began to speculate how they might be used effectively in all types of teaching and learning situations. The combination of high-speed processing and large-scale storage could enable teaching, testing, and remediation programmed to create machine-based tutors. Furthermore, the tutors could be individualized to allow learners to proceed through a set of instructional materials at their own pace. Such speculations gave rise to predictions of dramatic changes in teaching and enormous improvements in learning.

The development of individualized tutors promised a restructuring of educational institutions at all levels, from pre-school to continuing occupational and professional certification. Students would no longer be grouped in grades corresponding to their age cohort. Rather, they would be assigned to learning groups based on their mastery of relevant knowledge and appropriate skills. Enthusiasm abounded for educational reform that promised gains in teaching and learning that had never been accomplished in all of history. Yet, more than 50 years later, that promised reform is finally gathering momentum. What accounts for the delay? The answer to that question is complicated, for it involves three separate but interrelated lines of reasoning: competing interests, resistance to change, and a lack of knowledge and understanding of what constitutes effective teaching and learning.

Competing Interests

The late Michael Mahoney, professor of history at Princeton University, delivered a lecture at King's College in London in 2004 in which he identified seven communities of practice that influenced the early development of computers. Each had its own history that predated computers, and each developed its own hardware, software, and unique applications. Mahoney preferred to speak of the histories of these developments, rather than a single history of computing. The seven communities of practice were:

1. Scientists and engineers for whom the first computers had been developed
2. Commercial data processing companies
3. Corporate and business managers
4. Manufacturing and production control managers
5. Communication industry managers
6. Military command and control officers
7. Computer scientists who made the computers operate with increased efficiency

Those who are concerned with teaching and learning in schools, colleges, and elsewhere are noticeably absent from this list of communities of practice that exerted considerable influence on the direction of computer applications for the past half century. It is simply and clearly the case that other priorities took precedence, and resources and efforts were not directed to education, one of the major institutions in society.

Of the seven communities of practice identified by Mahoney, only commercial data processing produced software that was subsequently distributed to many schools. The systems developed for the other six communities had little or no relevance for the activities and functions of education. Data processing companies needed programs for financial record keeping, text document production, and file management. This software was reasonably well developed by the time personal computers reached the market in the 1980s. Hardware manufacturers, particularly IBM and Apple, were eager to capture the education market. Both launched substantial hardware and software donation and discount programs for schools, and most of the software programs were spreadsheets, word processors, and operating systems that facilitated file management, exactly the major programs that had been developed for the data processing industry. The problem was that these applications had limited utility in schools. The software that was readily available to schools was developed for other markets, and many teachers struggled with only limited success to use these programs in their classrooms. The education market was not sufficient for either the public or the private sector to invest substantial resources to create useable software. Furthermore, even if the resources were available at that time, no one sufficiently understood then what kinds of software should be developed.

Resistance to Change

A second contributing factor to the delay in the development of ICTs in education was the initial resistance from the teaching professionals, including the teachers' unions. Many of the enthusiastic predictions that accompanied the introduction of both large-scale digital and personal computers forecasted the development of robotic tutors that would eventually replace the classroom teacher. Such predictions spread alarm among the profession, and very few, if any, attempts were made to allay these fears. More realistic and less ominous predictions would have recognized that the functions teachers perform would gradually change over time as the use of ICTs enabled new kinds of teaching and learning activities.

The role of the teacher would gradually transform from knowledge distributor to one of guiding individuals or small groups of students through a series of increasingly advanced mastery exercises mediated through computer-based tutorials as general distributor of the same body of knowledge to all students simultaneously. Rather than replacing teachers with computers, ICTs would facilitate more effective computer-aided teaching and learning. Also, one school of thought places more emphasis on teaching students the skills of problem solving, including accessing and using information, rather than memorizing a specific body of knowledge, which is quickly forgotten after the next standardized achievement test has been completed. This transition is sometimes described as moving the teacher from the role of the "sage on the stage" to that of the "guide on the side."

Despite the fact that individual teachers and school administrators enthusiastically embraced the challenge of the changing role of the classroom teacher, many other members of the profession expressed concerns for their job security; some of the union leaders were most vocal in organizing and expressing this alarm. The reaction of the unions raises a number of interesting questions about the status of the teaching profession. Max Weber, the German sociologist discussed in Chapter 2, describes a profession as a collection of individuals who practice service delivery to clients according to a code of ethics. The behavior of professionals is self-regulated; if questioned, it is referred to a central regulatory body for adjudication. Classic examples of professions include medical doctors and lawyers.

Professions are frequently contrasted with and considered antithetical to occupations represented by labor unions, for which collective bargaining and job security are central characteristics. Disagreements and conflicts between workers and managers are common in labor relations. In many fields, the lines of distinction between professions and unions have blurred, and this is true of the teaching profession at both the school and college levels. Witness the recent growth of unions among professors. The industrial conflict perspective was certainly evident in the position of teachers' unions with respect to the introduction of ICTs in schools. This situation is gradually changing, but covert opposition to technology in classrooms persists.

Understanding Effective Teaching and Learning

The third and probably most important factor that has delayed the effective use of ICTs in education thus far is the lack of understanding about what constitutes effective teaching and learning. As noted in Chapter 1, programming computers to accomplish a task whose components are known is a relatively straightforward matter. For example, programming a computer to keep the records current in an insurance company consists of writing instructions to post incoming premium receipts, outgoing claim payments, and changes in risk classes and rates for various contingencies. Each step of the processes can be specified and thus stated in an algorithm and in turn programmed for computer execution. Some programs can be so complex and have so many contingencies that the programming is always being updated. An international air traffic control system is an example of such complexity that the programs are constantly upgraded to account for such factors as increases in volume and routes of flights, changes in weather patterns, the operating characteristics of new aircraft, or even suspected terrorist attacks.

In the case of using ICTs for teaching and learning, little clarity exists in designing the appropriate algorithm. When one considers the magnitude of the tasks, this lack of clarity is quite understandable. "Teaching and learning" is a convenient phrase to describe a wide variety of activities, including all the subjects at all stages of development, all the styles of learning of the students, and all the dynamics of the interactions between the teachers and the learners. All the cells in this multi-dimensional matrix will produce an almost infinite number of algorithms, each representing a possibly unique computer program. But the problem is even more intractable, for one would need to know for each step in the learning process how mastery is achieved. This approach is impractical, but the perspective does help in understanding why progress in the implementation of ICTs in education has been so slow.

Innovation in Education

Thus far, this discussion may give the impression that very little has been accomplished in creating effective uses of ICTs in teaching and learning. That impression needs to be corrected, for a plethora of promising activities are underway in research laboratories and educational institutions around the globe. First, in recent years, more resources have become available from both public and private sources to support ICT developments in teaching and learning. Second, the earlier resistance among teachers is abating, in part because digital natives are entering the teaching profession, replacing the older immigrants. These younger teachers are more adept and enthusiastic about employing ICTs in their classrooms. And third, recent advances in cognitive science provide a theoretical foundation for creating the kind of ICT-based tutors that encompass instruction, assessment, and remediation.

There are so many interesting and innovative projects that only a few exemplary ones can be discussed here to give some sense of the variety of approaches that are being explored. Perhaps the longest running project is Programmed Logic for Automated Teaching Operations (PLATO). Originally developed at the University of Illinois Urbana–Champaign in the early 1960s, the PLATO system drew upon prior work to develop teaching machines using psychologist B. F. Skinner's concept of programmed instruction. PLATO includes roles for students, instructors, and authors in creating, learning, and assessing student proficiency in a wide variety of subjects. When first introduced, PLATO was a major advance over the earlier and boring drill, practice, and test programs. After 45 years of development and expansion, PLATO Learning is now a NASDAQ-traded public corporation offering kindergarten through adult-level managed course content around the world via local area networks, CD-ROMs, and the Internet. In 2012 PLATO Learning changed its name to Edmentum to reflect its expanded perspectives and operations. The company reports that more than 1 million students and 65,000 teachers and administrators log onto the system every day.

The development and use of online courses in schools has grown in recent years. Many school districts hope to reduce operating expenses by increasing the number of students who take courses or parts of courses online. In the 2009–2010 school year, 2 million students or 4 percent of the national student body were estimated to have taken online courses in elementary, middle, and high schools. By 2020, all high school students are expected to be taking at least one course online. There is a great deal of school-based activity developing online curricula with local district support. Staples High School in Westport, Connecticut, for example, has developed online curricula for a series of mathematics courses. Similar curricula are being developed in mathematics and other subjects in many high schools around the country, and many are putting them in the public domain and thus making them available at no charge to other schools and districts. These materials also find their way into the programs of home schooling parents who are supervising their own children's education.

Many charter schools include online courses in their programs. Some contract with a commercial supplier of online curricula, such as K12, Inc. Many of these courses complement regular class curricula. However, 18 states now authorize virtual charter schools in which students take all courses online and never physically attend classes. Many educators believe that virtual schools are a useful future option for school systems, particularly to alleviate crowded classrooms. Students can study at any time and advance at their own pace. As with charter schools generally, the financial arrangements can cause controversy. School districts, boards, and teachers' unions have sued in court to prohibit virtual charter schools on the grounds that they take away too many resources from the existing programs. They also object to contracting out to for-profit suppliers for services that traditionally have been supplied by in-house professional staff. Some concerns have been voiced that virtual charter students do not benefit from interpersonal interactions with teachers and other students. Some of the previously mentioned teacher hostility toward educational technology likely fuels part of these controversies. Equitable resolutions of these issues remain elusive.

Another category of promising uses of ICTs in teaching and learning involves systems analysis. In these programs computer modeling simulates the dynamic relationships among different components of a real-world phenomenon. For example, models have been created and used successfully in teaching and exploring the laws of motion, stock market behaviors, corporate management, international relations, and climate change. Students build models or modify existing ones and then observe the dynamics of system changes over hypothetical or simulated time periods. Learning is enhanced when students can alter and observe the interactions of system components over simulated time. This theoretical field is known as *systems thinking*, and it was originally developed by Professor Jay Forrester at the MIT Sloan School of Management. In 1985, the late Barry Richmond, a student of Forrester and at that time a professor at the Tuck School of Business at Dartmouth College, established High Performance Systems, a company that developed the program STELLA, which implements the systems thinking perspective. Following Richmond's early death, the company reorganized with colleagues and family members under the name of iSee

Systems, and it is the current leader in producing systems thinking modeling software and promoting its use in education. Systems thinking is now used extensively in middle and high schools all over New England, and in the states of Arizona, California, and Oregon.

An interesting case relevant to the difficulties of using ICTs for teaching and learning in an international setting is the One Laptop per Child (OLPC) project. Established in 2005 by Nicholas Negroponte, who at that time was the director of the MIT Media Lab, the project was intended to design, produce, and distribute notebook computers to millions of children in developing nations around the world. Each child would have his or her own computer with a rechargeable battery. The software would allow each child to explore and engage in what Negroponte and his two MIT colleagues, Seymour Papert and Alan Kay, called *constructivist learning*. This approach emphasizes learning by doing and integrating prior knowledge with experience. OLPC planned to mass-produce millions of computers and expected that the price per unit would come down to $100.

A large number of U.S. corporations and foundations contributed to the project. However, the number of developing nations that agreed to participate was disappointingly low. Concerns were expressed that OLPC represented a neo-colonialist perspective, and many believed that investments in local schools and libraries would be more effective in developing the educational systems to expand local economies and labor forces. Distribution of the computers started in 2007, and by the end of 2010, only 2 million had been delivered to just four African nations, five Latin American nations, three in Asia, and a handful in Oceana. The numbers were a huge disappointment.

It is useful to speculate about the causes of the limited success of what initially appeared to be a most promising project. A number of factors may be relevant. First, when the project was announced, several U.S. computer manufacturers decided to attempt to produce machines at the same low cost and thus compete with OLPC in selling their machines to the educational systems in the developing nations. Second, a dispute arose over the operating system that would be installed on the computers. One supplier offered the software at no cost, but another supplier withdrew an offer of free software when it learned that its competitor would be involved in the project. Third, and perhaps most damaging, it became clear to many people that the technical and pedagogical support systems that schools and families would need to get any benefit from owning the computers were nonexistent in these developing countries. In addition, it would take years and enormous resources to create the infrastructure needed to realize the benefits of having access to the hardware and software. One would think that lesson had been learned when U.S. computer manufacturers flooded schools with personal computers back in the 1980s. Most of those PCs were vastly underutilized, if used at all.

Virtual field trips are another promising use of ICTs for teaching and learning. Excursions to a local museum, a factory, a college campus, or an historic site have always been a favorite school activity for children and teachers alike. Perhaps a large part of the positive reactions that field trips generate results from the break in the usual routine of the school schedule, but such activities do expose children to new places, perspectives, and ways of thinking. Although they are not yet used extensively in schools, virtual field trips potentially are a vast resource to complement traditional teaching and learning. It would not take a lot of effort from classroom teachers to use these resources effectively. Online films, videos, and websites can complement students' explorations of history, language, and culture studies.

Most large museums throughout the world now have online materials highlighting their collections. Teachers can guide their students through the Metropolitan Museum of Art in New York City or the Louvre in Paris. Students of ancient history can tour an online recreation of Rome at the height of the empire, and students of the U.S. Civil War can visit a simulated valley in Virginia that was at the heart of that conflict. Interactive instructional programs combine text, pictures, audio, and feedback to enhance the acquisition of competencies in second languages. The possibilities for expanding the dimensions of teaching and learning experiences through the simple accessing of the resources available online through ICTs appear to be limitless.

As mentioned earlier and perhaps most relevant to this discussion of teaching and learning and promising developments for the future are a series of research and development projects at Carnegie Mellon University under the direction of Professor John Anderson, a cognitive scientist. Anderson and his colleagues are developing what they call *intelligent tutoring systems* that both teach and assess student progress. These tutoring systems are based on detailed models of cognitive processing in teaching and assessment created by analyzing expert behaviors. Consequently, their development is time consuming and the costs are quite high. The tutoring systems were initially created for military services in training aircraft mechanics. The same structure was then applied in teaching several high school-level mathematics subjects. In recent years, the project has expanded to include physics, chemistry, genetics, geography, programming, and medical diagnoses. The work of Anderson and his colleagues is most promising, for it creates a machine-based tutor that is constructed on a cognitive model of the learning process. It does not attempt to apply existing software to enhance teaching and learning. Rather, it uses the processing power of ICTs to implement a theory of learning that provides stimuli, analyzes responses, and offers remediation. Although costly to develop, intelligent tutoring systems hold great promise for the future of teaching and learning in a wide variety of subjects and for students at all levels.

HIGHER EDUCATION

Today's colleges and universities around the globe are overwhelmed by numerous changes that cause them to revisit questions of their mission and methods. Escalating operating expenses, diminished financial resources, changes in the ages and aspirations of students, and the opportunities and problems accompanying the proliferation of ICTs all pose enormous challenges to those responsible for leading higher education into the twenty-first century. Professor Clayton M. Christensen, Harvard Business School, and Michael B. Horn, Executive Director, Innosight Institute, recently published an article describing the many difficulties facing higher education. Drawing on Christensen's earlier work on change in business organizations and schools, the article promotes the perspective of "disruptive innovation" to guide the revision of the mission and means of goal accomplishment for all colleges and universities. They call for a radical reformulation of these institutions that entails more effective means of using ICTs to create the foundation for lifelong learning patterns.

This section of Chapter 5 first briefly reviews the history of higher education in the U.S. from the perspective of the challenges and opportunities afforded by ICTs, and then it briefly discusses expanding activities in international higher education. The chapter then covers two additional topics: new formats of instruction and the changing foci of academic libraries.

Higher Education in the U.S.

Throughout the history of the U.S., the mission of higher education has undergone several fundamental transformations. During the colonial and Revolutionary War periods when the first colleges were established, the primary function of these institutions was the training of future clergy. The curriculum, therefore, was hardly practical; it consisted solely of the study of theology and the ancient languages of Greek, Latin, and Hebrew. The ecclesiastical affiliations and their influences on the curriculum lasted well into the twentieth century. Even today, both the Roman Catholic universities and some of the recently established evangelical Christian colleges provide theological components to their curricula for a substantial number of students.

Shortly after the Civil War, the U.S. Congress passed a series of laws granting the states parcels of federally owned land for the purpose of establishing colleges to teach and improve the practices of farming and mining. These land-grant colleges later added research and advisory services to the residents of their states. Thus were established the first instances of a practical curriculum, research, and community services in higher education. The land-grant colleges and universities initiated a trend that led to fundamental changes in higher education in the U.S. that persist today. The instructional and research programs gradually expanded from agriculture and mining to include new areas of engineering and science. As the mass production economy in the U.S. grew in the late nineteenth and early twentieth centuries, higher education took on the task of training future scientists, engineers, and managers for the rapidly expanding technology-based industries. Thus, the foundations were laid for the subsequent professional programs in a wide variety of fields. In all these programs, ICTs influence patterns of instruction and research.

It is interesting to note that the land grant institutions initiated programs of distance education to provide opportunities for students and adults to study continuously to improve their productivity. Consequently, distance education has a long tradition among those institutions. Correspondence courses and county agents supported the colleges' missions of off-campus education in their respective states. The University of Maryland and Pennsylvania State University became leaders in correspondence courses and degree programs at both the undergraduate and graduate levels. These programs were the precursors of the contemporary efforts to use ICTs to expand the substance and scope of education around the globe.

Prior to the dawning of the twentieth century, training in law and medicine took place in an assortment of diverse and independent apprenticeships and training schools of questionable quality. Although several universities, including Harvard, Yale, and Columbia, established graduate law schools in the early nineteenth century, apprenticeships and independent reading of law texts remained the most common means of entering the legal profession until the 1890s. With pressure from the then newly established American Bar Association, training for the legal profession was transformed into three-year university graduate-level programs.

Likewise, at the turn of the twentieth century, public concern grew over the declining quality of medical services in the U.S. and Canada. In 1910, Abraham Flexner, an educator and Carnegie Foundation staff member, authored a report under the auspices of the Carnegie Foundation for the Advancement of Teaching that detailed the deplorable state of medical training and called for extensive reforms. More than half of the existing medical schools then closed, and others became affiliated with graduate programs within universities and large training hospitals serving nearby communities. The formal attachment of law and medical training to universities was intended to emphasize the historical and scientific foundations of these professions. As with agriculture and mining colleges, the professional schools of law and medicine assumed the responsibilities of teaching, research, and community service. Universities then added a variety of other undergraduate and graduate professional training programs, including business administration, journalism, education, social work, librarianship, and more recently public policy in such areas as criminal justice, public health, and international relations.

Institutions of higher education now face shifts in the interests and educational objectives of their students. The proliferation of programs in colleges and universities signifies a trend away from the more traditional liberal arts and humanities subjects to the professional and vocational programs that prepare students directly for future employment. Apparently more students are interested in gaining competence in newer fields such as multi-media production that are expanding along with the increasing use of ICTs. The weakened economy beginning in 2008 has produced more concern among students about their future financial security. Appreciation of a liberal arts education has not disappeared in the academic world, nor is that likely to happen in the near future. However, the rising costs of higher education, and the attendant increases in tuition and fees, cause many students to enroll in the less expensive programs of two-year community colleges with the intention of possibly transferring later to a four-year program. The prevailing pattern of attending a four-year program immediately after high school and completing baccalaureate studies at the same school in four years is rapidly vanishing. Many students today transfer to several institutions en route to their degrees, and direct their efforts more toward earning credits, certificates, and degrees than on building a base of knowledge and skills to enable them to be effective lifelong learners. In addition to the changes in patterns of enrollment in higher education, more adults are beginning or returning to school, especially to the community colleges. The average age of all college students, full- or part-time, is now in the mid-twenties.

The changing interests of many students from traditional liberal arts undergraduate programs to a more vocational or professional orientation has caused alarm among many scholars, administrators, and business leaders. They worry that enrollment in courses and programs in the humanities and social sciences are decreasing and important elements of higher education are in danger of disappearing. Under the auspices of the American Academy of Arts and Sciences, the Commission on the Humanities and Social Sciences recently completed a review and published a report calling for the reestablishment of these scholarly disciplines in the teaching and research programs in all colleges and universities. Among other recommendations the report called for expanded funding resources in the humanities and social sciences to parallel those in the science, technology, engineering, and mathematics (STEM) disciplines.

To help offset rising operating expenses, many colleges and universities are turning to adjunct, or part-time, faculty. Adjuncts are hired on short-term contracts for a semester or academic year and receive a flat fee for their services, usually a fraction of a full-time faculty salary. The fringe benefits of retirement contributions, health insurance, and vacations that are routinely paid to full-time faculty are not usually available to adjuncts. Consequently, many adjuncts in large urban areas teach several courses during the same semester, spending much of their time traveling between campuses to meet their classes. An unfortunate consequence of this pattern is that students are unable to find faculty to interact with on an informal or even a scheduled basis. Colleges and universities can save a lot of money with adjunct faculty, but students receive limited guidance and must self-direct their academic progress. Increasingly, they turn to ICT-enabled learning. Many students use social networking to participate in online learning communities.

International Activities

The establishment of cooperative and collaborative programs involving a number of universities around the globe is an interesting and promising new activity in higher education. The combination of the Internet and cloud computing enables these international collaborations. An exemplary project underway in Europe since1999 could be a model for increasing international collaborations on a global scale. The project, known as the Bologna Process, with 45 European nations now participating, creates common standards for awarding three levels of academic degrees: bachelor's, master's, and doctorate. The intent is to foster intellectual growth and development among students and faculty by facilitating movement across institutions in all participating countries, known as the European Higher Education Area. The scope of the Bologna Process has expanded to include lifelong learning and professional teacher education.

When online courses and degree programs are fully implemented in the Bologna Process, for example, a student in Iceland will be able to take courses and receive credit or a degree from an institution in Ukraine, or a faculty member in Moldova can accept a remote appointment in Estonia. As one might expect, there is some opposition to this plan, particularly among the older faculty and more prestigious universities. However, the Bologna Process is advancing and seems destined to succeed eventually. If institutions of higher education in North America should create a similar consortium, it would greatly facilitate further academic progress and intellectual development for both students and faculty. Universities in the U.S. Midwest have expressed some interest in creating a local version of the Bologna Process.

Another promising development involves the creation of a consortium of higher education programs in the engineering profession. With leadership from Columbia University in New York City, a number of undergraduate and graduate engineering programs are expanding to include collaborative instructional and research activities. The institutions include universities in North American, India, and China. Students and faculty in participating institutions have an opportunity to visit and collaborate on a variety of projects addressing such issues as energy conservation, climate control, and sustainable construction. The projects involve the collaboration of private international corporations to enhance entrepreneurial activities among students and faculty.

Seminars and debates increasingly take place on a scheduled basis involving faculty and students from institutions of higher education in a variety of nations. Despite the substantial differences in local times, weekly seminars increasingly take place synchronously with participants in countries from most international time zones.

Changing Formats of Instruction

Turning now to alternate formats of ICT-enabled instruction, the use of online materials in courses takes four forms. First, online materials are increasingly used in traditional courses as supplementary resources. Professors used to put these materials on reserve in the campus library, or they contracted with local printing houses to produce packets of reprints for students to purchase. Most of these materials are now available online, and they can include websites that allow interactive learning, multi-media components, and mashups combining displays from a variety of databases. The second format is hybrid courses, which combine elements of traditional and online courses. Students meet in regularly scheduled classes, and they participate in online discussions between face-to-face sessions. The third format is the completely online course in which the students never meet with the professor or the other students. All lectures, readings, discussions, and examinations are online. The fourth format is the "flipped" course, a novel innovation that many see as a prototype for instruction in the future. In this format students complete all the assigned reading and assignments online in advance of the class meeting. Then the instructor uses the class time to interact with the students discussing content, answering questions, or helping students individually or in group work on assigned problems or projects. In the flipped format, the activities in the classroom or lecture hall shift from content delivery to a focus on the processes of learning. The -Babson Survey Research Group conducts annual surveys of online uses in the 2,500 U.S. institutions of higher education. In 2012 more than 6 million students were enrolled in some format of online courses. The percentage of higher education students taking online courses has increased from 9.6 percent in 2002 to 29.3 percent in 2012. Online courses are clearly the fastest growing activity in colleges and universities.

All the changes in the ages, enrollment patterns, and subject preferences are giving rise to new practices in pedagogy and promoting interesting innovations in higher education. Fifty years ago, most college-level courses were taught by a senior professor in a large lecture hall format with subsequent discussion sessions led by a graduate student. That pattern is gradually giving way to smaller, interactive classes using mobile devices for communication among students and instructors. Two promising examples of flipped courses are discussed here. The first is described in a recent book by professors Douglas Thomas and John Seely Brown at the University of Southern California. They created an innovation intended to replace the large lecture hall format, and they call it "a new culture of learning." It engages students actively in collective learning exercises. Students work in small groups on projects that are guided by faculty. The professors structure the course and facilitate the learning experiences. No one person is the dispenser of knowledge. Thomas and Brown report that students are enthusiastic about this style of learning. Furthermore, there is ample evidence of greater student involvement and knowledge acquisition and retention. This project has the potential of transforming teaching and learning at many different levels, and it is especially appropriate for students who will be functioning in a rapidly changing, technology-rich world. It is mentioned here because ICTs play a major role in facilitating student and instructor interaction and communication during the course. This is an interesting example of a pedagogical innovation enabled by ICTs. It is not the consequence of educators looking for a way to use computers.

The second example of a flipped course was developed in an introductory physics course for pre-medical majors at Harvard University taught by Professor Eric Mazur, an internationally recognized senior scientist and researcher. For many years Mazur had delivered lectures to large classes of students, and he became increasingly discouraged by repeated demonstrations that students did not retain the materials he covered in his very popular lectures. Mazur refers to a comical YouTube video featuring Father Guido Sarducci promoting the idea of a five-minute university. In the video Sarducci claims that any student can learn in five minutes everything they will remember five years after they graduate. Tuition is only $20 and includes a diploma and a photograph taken in a rented cap and gown. Mazur's course now employs what he calls *peer instruction*. The essence of this approach is that large lecture courses are taught promoting interactions among the students that are monitored by the professor. After a short lecture on a relevant topic all students record their answers to a multiple-choice question on any of a variety of mobile communication devices. The responses are immediately tallied and projected on a large screen for all to see. The students then engage in a discussion on the topic, producing peer instruction. Invariably a second polling of student responses produces a greater number of correct responses. By actively engaging with the topic in the small group student learning and retention is enhanced. Also, the professor gets immediate feedback for improving the short lecture content.

The largest provider of online education is the University of Phoenix, a private for-profit institution of higher learning that offers online courses and degree programs. It enrolls more than 500,000 undergraduate and graduate students, the largest of any higher education institution in North America. The University of Phoenix has come under scrutiny and criticism for what some charge are unscrupulous recruiting practices, particularly those efforts to recruit students who have access to federal funds to support education foillowing military service. Many colleges and universities are also expanding their own online offerings. California State University–Northridge increased its program of online courses from five courses 30 years ago to more than 300 in 2012, and it continues to expand its offerings to a growing number of students at all levels of post-secondary education. Stanford University now offers a master of engineering degree program that is completely online. In Japan, an online bachelor's degree program successfully competes with the traditional, highly sought-after universities in that country. Recently, Kaplan, the renowned test preparation company, collaborated with *Newsweek* magazine to offer an online master of business administration program.

The use of simulations and games in higher education has a long history dating back at least to the use of war games in training military leaders at service academies. Role-playing games were used widely with paper message exchanges to simulate diplomatic communications in schools of international relations. And simulations of the behavior of corporate executives served as common training vehicles in programs of business administration. Today, many institutions use games in their programs for such diverse audiences as journalists, military officers, and firefighters. However, the creation of virtual or micro worlds has increased substantially with ICTs. The most widely used and consequently the best-known virtual world is Second Life (SL). Created in 2003 by Linden Research, Inc., SL is a free, three-dimensional world where members can meet friends, conduct business, and share knowledge and experiences. Linden reports that more than 18 million adults participate worldwide. There is an SL for adolescents between the ages of 13 and 17. Members create an avatar to represent themselves when they are "in world." The avatars can take a human, animal, or object form, or any combination thereof. SL has its own currency, Linden dollars, allowing members to conduct a variety of businesses, buying and selling products, services, and land. Profits can be converted to U.S. dollars, and there are reports of SL entrepreneurs earning in excess of $1 million in a year. It is interesting to speculate if, how, and where taxes might be paid on such incomes.

Many people and institutions found SL to be an excellent locus for educational activities. Language instruction is the most common application, followed by fine and performing arts and religious instruction. Virtual science laboratories for assigned problems are a popular SL activity. They allow students to engage in collaborative learning and tinkering experiences, a form of learning that John Seely Brown, the renowned cognitive and computer scientist mentioned earlier, says used to be an important educational experience. Many colleges and universities have established virtual campuses in SL, and avatar faculty members conduct courses. An increasing number of professional business, law, journalism, and nursing schools offer degree programs within SL. Some SL campuses supplement the efforts of admissions offices by offering virtual tours of both real-world facilities and the various academic and extracurricular programs. For example, Pennsylvania State University uses SL to enhance its academic advising program for students on its 28 campuses throughout the state. Advisors are available at SL office sites at pre-set and announced hours to assist real-world students in their course and program planning. Some colleges and universities have recently aborted their activities in SL. It remains unclear whether these withdrawals are a consequence of reduced budgets in higher education generally or if there is some disenchantment concerning the efficacy of these kinds of distance learning experiences.

An interesting recent application of online continuing education is occurring among some professional schools as a means of strengthening relationships with their alumni. The schools offer programs of continuing education and in some instances of renewable degrees with online courses. The concept of *renewable degrees* acknowledges that the rapidly expanding body of knowledge and skills necessary to remain proficient in many professions today will require additional training, and the schools wish to capitalize on this opportunity to offer such training and certification. Such programs can also increase donations and revenue streams. A graduate of a professional school is more likely to remain a regular contributor to the school's fund-raising efforts if there is a continuing educational bond. Of course, the value of renewable degrees would have to gain widespread acceptance in the appropriate practicing professions.

Another interesting and promising development in higher education that is afforded by ICTs concerns the open courseware movement (OCW). First started by MIT in 2000, OCW puts all the materials for a course online for anyone to access and download at no charge. The materials include syllabi, lecture notes, tutorials, examinations, and videos. MIT has now posted these materials for all undergraduate and graduate courses. It has also developed a number of courses for high school curricula and included them in OCW. Use of the MIT materials has spread around the globe, and people in India and China are particularly active in accessing them. Inspired by MIT's contributions to worldwide education, a number of other colleges and universities have formed the Online Courseware Worldwide Consortium (OCWC) on the Web. Twenty-five additional U.S. colleges and universities have joined the consortium with varying commitments to post the materials for all or nearly all their course offerings. More than 50 countries worldwide are now members of the OCWC, thus creating a vast collection of free materials for higher education. Accurate data on the uses of these materials are not yet available, but they promise to tell a potentially inspiring story of the advancement of global higher education.

In 2007, Apple released iTunes U, which provides a similar service to OCWC. This free software platform facilitates the uploading and downloading of files, including texts, lectures, seminars, language laboratories, athletic events, and even campus tours. Colleges and universities can use iTunes U to distribute their own materials on their campuses and worldwide; iTunes U is available in the U.S., Canada, the U.K., Ireland, Australia, and New Zealand. Many leading institutions of higher education around the world use iTunes U to place online the materials supporting many of their courses. Both OCWC and iTunes U are facilitating an open source movement in higher education that provides widespread access to an enormous corpus of instructional materials to anyone, at any time, from any location. Of course, the creation of an effective learning experience will require more than access to course materials. Nevertheless, OCWC is a first step in the direction of creating global access to higher educational instruction.

Massive and global online access to higher education courses has expanded recently in both content and format. Massively Open Online Courses (MOOCs) are touted by many advocates as the vehicle that will promote vast changes in higher education worldwide. MOOCs are organized as online courses; and they include video lectures, supplementary readings, homework assignments, and examinations. Students who participate in MOOCs can complete courses from anywhere in the world and receive acknowledgement. In some cases the acknowledgement is a certificate or badge of completion, and in others the students receive credit that may qualify toward a degree.

Columbia University initiated an early form of a MOOC in the Fathom Program in 2003. In 2006, Stanford, Yale, and Oxford launched a consortium, AllLearn, which eventually offered 110 courses to more than 10,000 students in 70 countries. Neither of these programs survived, but they provided important experience that was helpful in subsequent efforts. In 2012, three large programs launched employing MOOCs to serve many students around the globe. One program, edX, was a collaborative effort on the part of Harvard and MIT. The other two programs, Udacity and Coursera, were launched from Stanford. Important similarities and differences exist among the three programs. For example, edX is a non-profit organization; and Udacity and Coursera are for-profit companies.

The edX program drew on the earlier efforts by both Harvard and MIT to place materials online for all undergraduate courses, including lectures, examinations, and supplementary readings, thus providing open access to all the instructional materials to any person anywhere. This in itself was a radical departure from past practices when most professors viewed their syllabi as private property. The edX program then created MOOCs in a variety of disciplines by combining the already existing online materials with the structure of a course with assignments and examinations. Software grades the assignments and tests. For a small fee people who complete the course satisfactorily can receive a certificate. As stated earlier, edX is a non-profit organization. Harvard and MIT each contributed $30 million to launch the project. Hopefully, a viable plan for sustaining the project can be implemented. By the spring semester of 2013, edX planned to offer 10 courses, and the project had expanded to include The University of Texas and San Jose State University.

The second MOOC, Udacity, was also launched in 2012 by Stanford University faculty Sebastian Thrun, David Stavens, and Mike Sokolsky. They offered an introductory artificial intelligence course that was derived from a similar course taught at Stanford in the previous academic year. The response to the Udacity course was overwhelming; more than 150,000 students registered worldwide. Building on that experience and subsequent collaborations with other universities, the trio then launched with venture capital support the for-profit company Udacity. The name symbolizes the intent of the company as " . . . we are audacious for you, the student." Udacity has collaborated with a number of commercial companies including Google, Microsoft, Autodesk, and Wolfram Research to offer online courses for employees. When the course is successfully completed, the employers receive an updated resume on the employee. Udacity has also collaborated with nearby San Jose State University to offer a number of online mathematics and statistics courses.

The largest of the three MOOCS, Coursera, was launched as a non-profit corporation in 2012, with start-up funds provided by venture capitalists. Two professors, Andrew NG and Daphne Koller in the Computer Science Department at Stanford University, originated Coursera, but it quickly expanded beyond computer science to include healthcare, medicine, biology, humanities, and social sciences. Early collaborators included The University of Michigan, Princeton University, and University of Pennsylvania. By the spring of 2013, Coursera included 62 colleges and universities offering MOOCs to 2.8 million users. Quizzes are graded either by students or software, and final examinations can be proctored for a fee, which lends greater credibility to certificates. The American Council on Education works with Coursera to determine appropriate levels of credit for different courses. Coursera is exploring a variety of ways to produce revenue to insure future operations, such as selling information to potential employers.

MOOCs could potentially provide dramatic changes in the structure and functioning of higher education. Advocates are labeling MOOCs as the University of the 21st Century. If a sufficient number of institutions jointly participate in a consortium of MOOCs, a pattern similar to the Bologna Process mentioned earlier might develop in the U.S. or even North America. On the other hand, many critics of MOOCs maintain that an important part of the higher educational experience is the interpersonal interaction between and among students and faculty. For example, a number of liberal arts colleges have decided not to participate in Coursera, for they wish to maintain and nurture further the culture of the small residential institution. Given the phenomenal growth of MOOCs in just a few years, there clearly is a demand for remote learning experiences. Initially, it seemed that higher education administrators were enthusiastic about the growth of MOOCs, for they believed online courses would address some of the problems of cost containment and greater student access. On the other hand, many faculty worried that MOOCs would destroy the relationship between faculty and students, and they also express concerns that jobs may be lost to the technology. Faculty at Duke and San Jose State Universities as well as Amherst College has voiced strong opposition to MOOCs on their campuses. It seems likely that in the near future a combination of traditional and MOOC structures may prevail.

An additional aspect of the changing formats of higher education involves the creation, distribution, and use of textbooks and related materials. For years, the commercial textbook market was an important and lucrative ancillary industry supporting higher education. Publishing houses attempted to provide a single text that would cover the topics included in most typical undergraduate and graduate courses. Faculty found one major text per course to be a promising cost-saving alternative for the students, and a few professors became wealthy from royalty payments when the texts they authored enjoyed widespread adoptions and sales. However, as the costs of producing texts continued to soar in the 1980s, students and faculty started searching for a less expensive means to provide these materials. The used textbook market was an attempt in this direction, but the cost savings to students were not sufficient. The Internet provided a number of useful alternatives. Online publication was one obvious possibility, but eliminating the costs of printing and distribution did not reduce the prices sufficiently, and many students and faculty reported that they were uncomfortable reading texts on the screens of computers or mobile devices. Some publishers and higher education institutions are exploring site licenses as an alternative. Under this arrangement, the college or university purchases a license from the publisher, and the textbook, or portions thereof, can then be downloaded, saved, read, and printed by the students enrolled in the course. The cost to the student is greatly reduced under this optional licensing and print-on-demand arrangement. Depending on the fee paid by the college or university, this might be a viable alternative for publishing houses.

Another variation of this option is the customized text. In this case, a publisher becomes the provider of a large amount of online materials relevant for instructional purposes in a particular subject. Many of these materials may already be online through other projects which scan books and journals, making them publicly available. Faculty members select those materials suitable for their courses, and students download them as in the site license option just described. The authors who now write textbooks would be hired to produce or update the online materials for a fee or an advance on royalties of future use. The providers of these services would specialize in certain topics or areas, such as economics, U.S. history, or chemistry, much as they have for publishers of textbooks in the past. It will take some time to accumulate the extensive files of relevant texts, and it might be prudent to start with materials relevant to large undergraduate introductory courses. This use of ICTs could drastically change the business of textbook publishing, but it also has the potential of expanding the scope and upgrading the quality of the materials used in college and university instruction.

Scientific Research

The widespread distribution and use of ICTs in higher education has greatly impacted how research is funded, conducted, and supported in colleges and universities, especially in North America. As noted earlier, research is widely accepted as one of the major functions of higher education. With the establishment of the National Science Foundation (NSF) following World War II and the expansion of the National Insititute of Health (NIH), the federal government became the major source of support of U.S. academic-based scientific research. It has recently played a major role in the creation and expansion of resources to support research, especially in the disciplines of computer science, applied mathematics, physics, chemistry, engineering, and biology. The establishment of several supercomputer centers with NSF support now provides virtually unlimited processing capabilities to almost all academic researchers. Just 10 years ago, such computational power was limited to those people affiliated with just a handful of university-based research facilities. Research projects that use large-scale databases with extensive computation are now readily available to all scholars who have access to the Internet. Data analyses can now be completed on supercomputers and activated remotely from campuses located anywhere in the network. These capabilities allow scholars in all fields to explore new and expand existing investigations, including such complex topics as climate change, cosmology, archaeology, and demographic changes in diverse populations. The list of new scholarly inquiries is almost limitless.

For the foreseeable future, the finances to address the problems and opportunities that are enabled by ICTs will be severely constrained. With the world struggling to recover from a major economic downturn and avoid similar collapses, finding additional financial resources to take full advantage of the many opportunities mentioned in this chapter will be difficult. In the U.S., most state and local governments are in financial crises, and they are cutting back on all budgets. Especially hard hit are the budgets of many community colleges, and this takes place just as they are experiencing an increase in enrollments. Land-grant institutions are reporting record levels of tuition and fee increases, and students are protesting vigorously. At the University of California, a recent commission report recommended a three-year undergraduate program, an expansion of online courses and programs, and an increase in the number of students transferring from community colleges to universities. Some private institutions of higher education are experiencing as much as a 30 percent decline in their endowments, and financial aid to students is in peril. Despite warnings from leaders in both the public and private sectors that education remains a key investment in our future economic well-being, one must worry that the available resources will not be adequate.

6 Knowledge

CHANGING PATTERNS

As mentioned in Chapter 1, the late Daniel Bell published in 1973 his description of the post-industrial society, a prescient analysis of the dynamics of social, political, and economic change. He posited that the post-industrial society was characterized by two major axes of social structure: technology and knowledge. This chapter focuses on knowledge. It examines those areas in which information communication technologies (ICTs) are changing the patterns of creation, dissemination, and use of knowledge.

People have struggled with the definition of *knowledge* as far back as ancient Greek society. Contemporary philosophers and social scientists continue to search for an adequate definition and useful distinctions between and among the related concepts of *data*, *information*, *knowledge*, and *wisdom*. Rather than attempt to provide another set of definitions, this chapter focuses on the evidence that ICTs affect the processes by which data, information, knowledge, and wisdom are created, shared, and used. In his analyses, Bell proposed a useful metaphor to comprehend some aspects of these relationships. He compared the relationships among knowledge, information, and data to the format of a book. Knowledge corresponds to the content of a book, information is analogous to the subject index, and data serve a function similar to the name index. Knowledge reflects an understanding of the content. The subject index is a collection of headings that guides exploration of the content by topic. And the name index is ordered alphabetically and provides easy access to the contents via the names of people.

Before starting this inquiry, the scope must be limited, for the enormity and complexity of including all human knowledge make it impossible to cover in this limited investigation. Therefore, the analyses presented in this chapter are limited to scholarly, scientific, and technical knowledge in the physical and social sciences and the humanities. Before looking at how ICTs affect each of these aspects of knowledge production, it is necessary to note several ongoing patterns of change in the world of scholarship at large that are enabled by ICTs. Three interrelated topics are relevant: the growth of big science, the increasing practice of collaboration, and the expansion of interdisciplinary projects.

Big Science

The first pattern is the ongoing expansion of what is frequently called *big science*. The Manhattan Project was the first large scientific endeavor to earn that label. The project ran from 1942 to 1946 and produced the atomic bomb, credited with bringing World War II to a halt. A joint effort among scientists in the U.S., the U.K., and Canada, the project's work was distributed over 30 sites. At its height, the Manhattan Project employed more than 130,000 persons, and in present-day dollars, it expended more than $22 billion. The label *big science* simultaneously refers to the enormous costs, the large staff, the expensive equipment, and the importance of the product or products. Examples of current-day big science projects include the Hubble Space Telescope, the Large Hadron Collider, the International Space Station, the Human Genome Project, monitoring and predicting earthquakes around the world, and the global tracking of changing weather patterns.

The number of big science projects has increased in the past 60 years and are no longer limited to the physical sciences. Projects involving large databases are now common in both the social sciences and the humanities. The decreasing costs of data storage and the proliferation of cloud computing encourage the creation of large databases that include repeated measurements of the same phenomena or processes over time. Many of these large databases are designed to permit ongoing longitudinal studies of continuing phenomena for indefinite periods of time. A well-known, early longitudinal study tracked 825 men and women graduates of Harvard and Radcliffe Colleges for a total of 65 years and revealed interesting patterns of adult development, including measures of health and happiness. At the present time, hundreds of longitudinal studies are in various stages of development and implementation. In the humanities, many proposals and projects are underway to create huge digital databases of books, journals, papers, artworks, and more. Research projects in the humanities and the social sciences are also becoming big in terms of funds, staff, equipment, and the importance of results. A common theme in this chapter, and indeed throughout this book, is that such developments would not be feasible without ICTs.

Collaborations Among Researchers

A second trend in scholarly research that has become increasingly evident over the past 60 years is the collaboration among researchers. When one thinks of scientists of the past, the image that most frequently comes to mind is that of the lone investigator: Galileo observing and recording the movements of the planets, Newton contemplating the laws of motion, or Edison creating the incandescent light bulb. Historians of science point out that the image of the solo scientist is a gross oversimplification, and that most important advancements in science had predecessors or collaborators whose contributions have been lost or ignored over time. Nevertheless, the pattern of the solo scientist, or even a small group of collaborators, has given way to investigations conducted by teams of researchers working in a coordinated manner. Recent studies reveal a decreasing number of single-author scholarly books and journal articles and a corresponding increase in multiple-author publications. This change has been occurring gradually, but consistently, since World War II. Works by multiple authors are more common in the physical sciences, but they are now increasing in the social sciences and the humanities. The picture of the lone historian working with dusty old manuscripts in an archive is also beginning to fade. Team research requires coordination, communication, and resource sharing, sometimes at great distances and asynchronously. ICTs facilitate such collaborations regardless of time differences and geographical distances.

Interdisciplinary Research

The third trend in the patterns of scholarly research is the increasing number of projects that include team participants from different scientific and scholarly disciplines. The word that is frequently used to describe this pattern is *interdisciplinarity*, an awkward term, yet an accurate description of the concept. Many of the major projects underway currently require the perspectives of several disciplines to ensure success. For example, research in neuroscience may require the contributions of neurologists, cognitive psychologists, computer scientists, chemists, radiologists, and artificial intelligence experts. Evidence of increasing interdisciplinarity can be seen on many university campuses where new buildings house research efforts in various combinations of disciplines such as cosmology, neurobiology, materials science, climatology, artificial intelligence, and robotics.

In each of these examples, the physical plant of the new buildings facilitates interdisciplinary exchanges among scholars who previously were affiliated with and housed in separate departmental offices defined by their academic discipline. Today, they share a common space and contribute as team members of big science or humanities projects. It is interesting to speculate on the future physical layout of the campuses of higher education institutions. Will there still be buildings that house departments differentiated by traditional academic disciplines, or will there be a new physical plant that reflects the emerging interdisciplinarity? In the latter case, there may be a need for greater flexibility of the use of campus space to accommodate the ever-changing profiles of then-current topics of scientific and scholarly enquiry. The hallowed halls of ivy may have to give way to a greater change than most academics expected in the past, even though professors are not noted for their capacity to adjust to changes of any kind, let alone in their work environments. Questions have recently been raised as to whether professors will need private offices at all. Rather, they might work in common spaces designed to promote interdisciplinary collaboration. Such a change may not happen easily in the near future.

The remaining discussion in this chapter examines how ICTs are affecting knowledge production in three stages: creation, dissemination, and use. Correspondingly, it examines how ICTs enable the transition from data to information to knowledge. Of course, knowledge production is a complicated process. It is a continuous augmentation to and integration of new knowledge into previous and ongoing scholarly and scientific research. To characterize it as a linear progression is an obvious distortion. However, the three stages are useful as an informative or heuristic device for organizing an inquiry into the impact of ICTs.

KNOWLEDGE CREATION

ICTs enable three new developments in knowledge creation: the expansion of large-scale data sets; distributed computing; and the establishment of online reference facilities, such as encyclopedias, dictionaries, and thesauri. The phenomenal advances in computing speeds and storage capacities over the past 60 years have resulted in dramatic expansions in the kinds and amounts of data relevant to problems that scholars and scientists can now pursue. It is useful to distinguish among three kinds of data that are germane to research: numerical, textual, and symbolic. However, all three data types share the common feature of using the capacities of ICTs for rapid processing. Numerical data are a mainstay in scientific research, and textual data play a similar role for much research in the humanities. Symbolic data refer to different formats and procedures in various fields that have opened new domains for investigation, for example, simulations, modeling, and gaming. As mentioned in Chapter 5, scholars increasingly use simulations that rely heavily on both symbolic and numerical data for research, because investigators can now construct and run many models on their high-speed desktop computers. In fact, the use of complicated multiple variable, multiple level, statistical data analyses has become a common practice in the social sciences, and ICTs facilitate their widespread use. However, it is useful to recall that all three data formats are represented in storage devices as binary bit patterns. The differences are in the meaning of the various patterns as recognized and processed by the internal registers in the computers.

Expansion of Large-Scale Databases

Many of the big projects underway today would have been unthinkable in the recent past. To illustrate how the employment of large-scale data sets has expanded in the past 60 years, it is illuminating to look at the history of three social science research projects: the General Social Survey (GSS), the National Educational Longitudinal Study of 1988 (NELS88), and the American Memory Project (AMP). The GSS has been conducted since 1972 by the National Opinion Research Center (NORC), a social science institute affiliated with The University of Chicago. The GSS conducts personal interviews with nationally representative samples of U.S. adults. The sample sizes are usually 1,500, but they are occasionally larger, up to 4,000 adults, to allow in-depth exploration of certain topics with respondents in specific subpopulations, such as racial groups, seniors, or immigrants. In the early years, the survey was completed every year; more recently, it has been conducted every other year. Each survey contains what the GSS staff calls the core questions or variables. Responses to these demographic, behavioral, and attitudinal questions allow the tracking of social trends over time. In addition to the core topics, the surveys have covered a wide range of different behaviors and attitudes over the years, including political participation, religious activities, mental health, racial prejudice, gun control, medical care, and gender equity. The accumulated files of the GSS are now huge and are continuously growing. They include more than 5,000 variables and almost 2,000 analyses of responses to the same questions over time. Obviously, substantial processing capabilities are required to manage such large and perpetually expanding data files.

The GSS is further expanding the scope and extent of its activities in three significant ways. First, it is converting itself from a series of cross-sectional studies in which different people are sampled and interviewed in each survey to a longitudinal design in which the same people are contacted in successive panels of the project. The longitudinal design enables researchers to follow the same subjects over time, thereby enabling more precise measures of social change. However, the complexity of the data processing system required to manage the records of persons from successive surveys is more challenging and requires more computing power. A system must identify individual records and maintain the promised confidentiality of the responses of all participants. Furthermore, these matching operations are repeated each time new data are collected in the recurring surveys.

Second, the GSS has expanded its scope by collaborating with the International Social Survey Program (ISSP). The members of this organization are committed to conducting comparable surveys in their countries, thus permitting international comparisons of social change. The ISSP was first established in 1982 by the U.S., Australia, Great Britain, and Germany. It currently has more than 40 members. Under the auspices of the East Asian Social Survey, China, Japan, South Korea, and Taiwan also participate in the international coverage. This expansion of the concept embodied in the GSS is greatly facilitated with the exchange of data files and expertise by ICTs. Indeed, it is difficult to imagine this volume of international collaboration without ICTs.

Third, the GSS is continually expanding the capacity of people around the globe to access and use the now voluminous data files of NORC and its international collaborators. These people access the GSS files for both research and teaching purposes, and such uses are increasing each year. To facilitate the domestic distribution and use of its data files, the GSS collaborates with three major U.S. survey research centers: the Roper Center for Public Opinion at the University of Connecticut; the Inter-University Consortium for Political and Social Research at The University of Michigan; and the Survey Documentation and Analysis Archive at the University of California, Berkeley. For social scientists, these files are an exceptionally rich resource for studying various aspects of social change, and there are multiple means of accessing them including file downloads, CDs, and cloud processing.

The use of the GSS materials as a teaching resource is particularly innovative. Secondary schools and higher education institutions find the GSS data files useful in a variety of courses dealing with the methods of social science research and the substance of social change. Faculty report that working with data files documenting patterns of social change in behaviors and attitudes provides an effective way to engage students in the subject matter of their courses. Hypotheses formulation and testing assignments are easily accomplished with the GSS data and readily available statistical analysis programs. The increasing proliferation of ICTs makes it possible for such big projects as the GSS and its international collaborators to collect and disseminate such huge data resources for research and instructional purposes.

A second example of a big data project is NELS88. Sponsored by the U.S. Department of Education, NELS88 is a longitudinal study of a national sample of students. The first panel of this survey was conducted in 1988 when the students were in the eighth grade. Four successive waves of the study collected data on the same students as they progressed through high school and on to post-secondary education and into the labor force. By the time of the last wave in 2000, a huge amount of data had been collected from approximately 24,000 subjects documenting their backgrounds, activities, and educational and occupational aspirations and achievements. In addition, data were collected from samples of the students' parents, teachers, and school principals. As with the GSS, the data files from NELS88 have been widely distributed and used for both research and teaching purposes.

A third example of a large-scale database activity is the American Memory Project. Operating under the auspices of the U.S. Library of Congress, the AMP was launched in 1994. It aspires to collect and make available to the public at no charge a comprehensive digital record of U.S. history. With funding from Congress and many private sources, the collected digital files will record materials that were originally created in a variety of formats, including written and spoken words, still and motion images, prints, maps, recordings, sheet music and scores, and a wide variety of collected ephemera. The massive holdings of the Library of Congress will form the bulk of the data files. However, the project plan includes digitizing the holdings of many other libraries, museums, and private collections and adding their files to the corpus of the American Memory Project archive. There are already more than 100 thematic collections on such topics as political parties, education, religion, civil rights, crime, and technology. Undoubtedly, the creation, maintenance, and continuous expansion of this archive will be of great value to future historians. Furthermore, remote access to this resource will be available through ICTs.

The GSS, NELS88, and the AMP present just three examples of large-scale databases commonly used in social science and historical research and teaching. Virtually thousands of such large databases are available in all the scientific and humanitarian disciplines. Many are longitudinal, and data are continually being added as new observations are collected and integrated. The topics covered in these large databases range from anthropology to zoology and include all subjects in the sciences and humanities.

A recently announced joint project involving Google and Harvard University scientists provides another excellent example of a new scholarly endeavor that would be unthinkable without ICTs. The project, known as Culturomics, has created a database of more than 500 billion words drawn from 5.2 million books written in seven languages: English, French, German, Spanish, Hebrew, Russian, and Chinese. Many university and research libraries have scanned and digitized copies of the books. All the words and dates of publication have been entered into a huge searchable database. Most of the books date from 1800, but some go back as far as 1500. The books represent approximately 4 percent of all books ever published.

Users can either download the database for their own research, or employ a Google program, the Ngram Viewer, to search the database and identify how frequently specific words or phrases appeared in any year. The creators of the system claim that scholars can examine a wide range of historical and cultural questions with the database. For example, when did a particular word first appear, or how frequently did one historical person's name appear in contrast with another's? Google has made both the database and search and graphing program available to the general public free.

Distributed Computing

Distributed computing involves the creation and maintenance of large-scale databases, but it differs from the projects just discussed in that more people collect the data and are frequently invited to participate in compiling and analyzing the information. The large-scale collections described earlier are completed by one or several organizations, and the data are gathered by professionally trained field staff for the surveys or by librarians, archivists, and curators in the case of the American Memory Project. In distributed computing, more individuals contribute to the activity, and they are frequently located around the globe.

The Berkeley Open Infrastructure Networking Computing (BOINC) project is one of the largest distributed computing projects. Supported by the National Science Foundation and headquartered on the Berkeley campus of the University of California, BOINC is an open source program that enables volunteers to donate both their own time and the use of their personal computers to help in a variety of scientific investigations ranging from scanning the Milky Way for stellar streams, to working on cures for a variety of diseases, to studying global climate patterns, and even to predicting earthquakes. More than 50 BOINC-supported distributed computing investigations are ongoing. Perhaps the best-known project is the Search for Extra-Terrestrial Intelligence (SETI). Launched in 1999, SETI now involves 3 million amateur volunteers all over the world. They search the universe and beyond to detect radio pulses that may indicate some form of intelligence in outer space. They then forward the data describing their observations to SETI, where they are analyzed and added to the burgeoning database of astronomical phenomena. This project gained substantial public attention when the film *Contact* portrayed a fictional account of some of the experiences of those involved in the searches.

Online Reference Materials

Online reference materials include library card catalogs, periodical guides, dictionaries, thesauri, and encyclopedias. Not too many years ago, such standard reference materials were accessed by handling the physical objects in a library, office, or at home. Now these reference materials are increasingly available online. For example, almost all academic and community libraries now have their card catalogs online. These online catalogs not only save floor space, but they also allow persons remotely and simultaneously to search the records of every item in the library's collections. Many professors report that they have not visited their campus libraries for years. They search the card catalog on their computers, and other faculty and students can search at the same time. They can also access the catalogs of other libraries, both near and far, including even the largest library in the world, the U.S. Library of Congress.

In addition to catalogs, a vast array of other reference works is available online. It is estimated that there are now 800 online dictionaries and thesauri for 160 languages. Furthermore, a huge number of special collections are now available to remote users, both scholars and the general public. For example, the venerable Cambridge University special collection consisting of more than 7 million volumes of antiquarian and rare books and manuscripts has been scanned and is now available online for remote access. Among the many and varied reference materials involved in the creation of knowledge, the production and use of encyclopedias are the most dramatically affected by ICTs.

The number and size of various online encyclopedias keep increasing, and the topics covered encompass everything from the trivial to the esoteric. In the recent past, many homes and offices contained multi-volume sets of commercially printed encyclopedias used for a variety of reference tasks or school research assignments. Most of the publishers of those printed volumes have now ceased producing books; instead, they provide online access to their materials. The growth pattern of such reference works is beginning to approach what Vannever Bush described in his prescient 1945 article in the *Atlantic Monthly* as a *memex*. Bush, who served as head of the U.S. Office of Scientific Research and Development during World War II, envisaged the memex as a system in which all the knowledge in the world would eventually be stored on microfilm. User access to any subject would be linked by a system of cross-references. Today, World Wide Web sites have replaced the contents of Bush's microfilm files, and search engines scanning Web addresses now provide the cross-referencing functions. Of course, not all the knowledge of the world is included in the Web, but ICTs are certainly moving in that direction and at an increasing pace. The ancient Library of Alexandria purportedly contained copies of all then-existing books in the world, and ICTs may someday enable an online approximation of that objective.

The digitalization of so many different reference materials marks an important development that ICTs enable, and clearly makes it much easier for everyone to access information from their home, office, college room, and most recently on their smart phone. However, in terms of the impact on the processes of knowledge production, the creation of the free, multilanguage, collaborative, online encyclopedia Wikipedia is the major innovation. A *wiki* is an open access website to which anyone can contribute materials. Correspondingly, Wikipedia is an open-source, online encyclopedia consisting of the inputs of literally tens of thousands of volunteers. Anyone can contribute an article or edit an existing entry. Wikipedia was launched in 2001 by programmers Jimmy Wales and Larry Sanger. Operating as a nonprofit organization, Wikipedia aspires to become a free source of all knowledge in all the major languages of the world and now has more than 10 million articles in more than 280 languages. Only 160 full-time staff members are necessary for this massive undertaking. Almost all the work of creating, maintaining, and expanding this site is done by volunteer labor, which is one of two things that makes Wikipedia unique. The other is that Wikipedia is by far the largest and longest-running massive participant or crowd sourcing project ever undertaken.

Crowd sourcing, a concept created by journalist Jeff Howe in 2006, refers to what he observed when the general public was invited by Proctor & Gamble in a widely distributed Internet message to participate in a collective activity. The company requested the public to submit comments and critiques of existing products and suggestions for new ones and was overwhelmed by the number and value of the responses. Although Proctor

& Gamble had a large in-house capability for product enhancement and creation, the company found that the public response was, in essence, an extension of their research and development output. Many other corporations and organizations copied this model and found similar patterns of productive feedback from the public. Some offered monetary rewards for valuable inputs, but others found that many people were perfectly willing to provide input for the simple pleasure of making a contribution to a publicly available resource.

Crowd sourcing quickly became a new tool for many organizations to solicit input from the public to enhance their goal accomplishments. Using crowd sourcing to solicit input, Wikipedia quickly became known worldwide as a new and potentially valuable source of knowledge. However, as social historians are fond of pointing out, all innovations have earlier roots that portend a forthcoming change. The *Oxford English Dictionary* (*OED*), first published in 1894, benefited from the volunteer work of hundreds of amateur lexicographers. People have always made contributions to different organizations, but ICTs have made crowd sourcing possible at a mass level that could never have been attained earlier.

An interesting aside concerning crowd sourcing concerns the recent flood of digital photographs sent to law enforcement officials that made it possible in a short period of time to identify the two people who allegedly planted the bombs near the finish line in the 2013 Boston Marathon. The media called this "crowd sourced criminal identification." Outraged citizens contributed thousands of pictures to aid in the investigation, and all the involved law enforcement agencies appreciated the assistance. The volume of replies did produce some false identifications that were circulated by television and newspapers with negative consequences for both the reporters and some innocent people. Although the investigation was clearly aided by the ICT-enabled crowd sourcing, the down side was that some persons were temporarily implicated.

Wales and Sanger created the software that enabled anyone with an Internet connection to place an article on its free online encyclopedia or to edit an existing entry. Those involved in this project sympathized with several social movements that espoused open source software, free culture, and collective wisdom. Based on the successful development and widespread use of such software programs as the Linux operating system and the Mozilla browser, proponents in these movements believed that society would benefit most from the free and open creation of technology, information, and knowledge. Wikipedia was the implementation of the open philosophy in the form of an online encyclopedia. It is also an implementation of the spirit behind Creative Commons, a nonprofit organization devoted to promoting more widespread use of copyright-protected materials. It is interesting to note that Wales serves on the board of Creative Commons.

As mentioned earlier, Wikipedia is the largest crowd-sourcing project in the world. Originally, anyone could add inputs to Wikipedia, and very little, if any, editorial oversight was exercised. If any erroneous materials were inserted into Wikipedia, the collective wisdom of the crowd was expected eventually to come to the fore, achieving accuracy and clarity. This mode of creating an encyclopedia, whether printed or online, was a radical departure from past practice; encyclopedias, as well as dictionaries, had always relied on the judgments of a large body of subject matter experts. Sometimes, these experts were compensated for the efforts; frequently, as in the case of the *OED*, the experts contributed volunteer labors of love. As might be expected, the absence of any editorial oversight in the early days of Wikipedia produced many incorrect entries, and critics quickly emerged. In fact, Wikipedia became associated with many extensive and gross errors. Teachers and professors frequently warned their students not to use Wikipedia at all. In response to such concerns, Wikipedia established what it calls Arbitration Committees to review entries or edits that appear to generate conflicts. A small group of experts reviews the matter and may decide to ban certain people or groups from posting on that subject in the future.

Among many incidents, two are used here to illustrate the kind of events that Wikipedia's editorial policies can create. The first involved a series of articles in which the authors posted different accounts of the activities of a nationally recognized religious organization whose members included several prominent stars of the entertainment world. The alternate postings became increasingly strident, and Wikipedia became the battleground for the internal conflicts among current and former members of the religious organization. Recognizing that such exchanges were certainly contrary to the mission of Wikipedia, the Arbitration Committee imposed a ban on the people and organizations involved. The second incident was provoked by a story printed in *The New York Times* and then widely circulated on the global wire services. It incorrectly reported that a well-known New England college had adopted a policy prohibiting all students from using Wikipedia as a resource for any of their research projects. After a prominent alumnus publicly challenged the college president for interfering with a collective wisdom experiment, an investigation into the matter was ordered. Subsequent inquiry revealed that one professor had merely advised students to use Wikipedia with caution. However, the second story never caught up with the first, and Wikipedia's image as an error-ridden source was reinforced.

The phenomenal success of Wikipedia launched a tsunami of wiki-related activities, all under the auspices of the Wikimedia Foundation, whose mission is to make all the knowledge in the world available at no cost to all the peoples of the world. For the most part, it persists with its policy of unedited and crowd-sourced materials. The contents quickly expanded from the encyclopedia to include a dictionary, thesaurus, rhyming guide, phrase books, language statistics, etymologies, pronunciation guides, synonyms, antonyms, and translations. Wikiquote is a crowd-sourced online version of the well-known *Bartlett's Familiar Quotations*. All Wiki knowledge sources are available in English, and many are posted in hundreds of different languages.

In addition, the Wikimedia Foundation supports many other activities in pursuit of its overall mission. Wikisource is a document library of 150 textbooks, and Wikispecies is a directory of 224 articles containing species descriptions. Wikinews is similar to a blog of up-to-the-minute news reports, and it encourages the development of participatory journalism. Wikibooks, a 35,000-page repository of books containing educational materials, is intended for learners from pre-school to graduate levels. Wikiversity is a 12,500-page repository of course materials, papers, and lessons also intended for teaching and learning activities from pre-school to graduate levels. Wikicommons is a general-purpose repository of 6.5 million files. Lastly, Wikimania is a website used to organize and facilitate an annual international conference for all participants in any of the wiki-related activities.

The abundance of wiki activities has spawned many other online knowledge projects. For example, the online Encyclopedia of Life operates on the BOINC software platform mentioned earlier. It aspires to gather descriptive information about all recorded forms of life on Earth. Also, there is Medapedia, an online source of information on all aspects of medical practice. However, these projects differ from Wikipedia in two respects. First, they have more narrow coverage. The Encyclopedia of Life is certainly an ambitious undertaking, but it is not focused on collecting all the knowledge in the world. The second distinguishing characteristic is that these encyclopedias do engage experts to judge the accuracy of all entries. Some of the experts are reimbursed for their labors, and others serve as pro bono contributors to the effort.

The now well-known WikiLeaks and the Wikimedia Foundation are completely separate organizations. Originally, WikiLeaks embodied the open and crowd-source philosophies that are the hallmark of all other wiki activities. However, as WikiLeaks pursued its unique interest of posting on the Internet information and documents that government agencies, business corporations, and private individuals wished to keep secret, all connections between the two organizations were severed. WikiLeaks and its director, Julian Assange, are ensnarled in a number of legal issues concerning the theft and distribution of confidential documents. As these alleged violations have occurred in multiple legal jurisdictions, it will take some time before they are resolved, if at all. The activities of WikiLeaks pose complex and controversial issues. Assange and his colleagues are viewed by many supporters as champions of openness and transparency of all the activities of both government and private organizations. They believe firmly that the public is entitled to all information; when any information is kept secret, corruption inevitably follows. Others view publishing confidential information as an unethical and illegal act. Advocacy of transparency is not a new phenomenon; history is replete with accounts of conflicts among various segments of society. However, the visibility, scope, and consequences of publishing confidential and secret documents on the Internet are greatly magnified by the use of ICTs.

No discussion of the online reference systems would be complete without mention of two other online encyclopedias, Citizendium and Conservapedia. Both were created in protest over the policies and content of Wikipedia. Citizendium was established by Sanger, who is referred to by many as the co-founder of Wikipedia. Sanger was upset by the increasing appearance of erroneous materials in Wikipedia and proposed introducing an editorial review function. However, he and Wales disagreed, and Sanger went off on his own and created Citizendium with what he called "a gentle expert oversight" of the content. Citizendium has yet to achieve the rate of growth of the Wiki Empire. Conservapedia was created in 2006 by Andrew Schlafly, a Harvard University Law School graduate and son of the conservative political activist Phyllis Schlafly. The impetus for its launch was a concern about the perceived liberal bias in the content of Wikipedia. Reportedly, the precipitating event was Andrew Schlafly's objection to Wikipedia's use of the increasingly accepted designation of Common Era (CE) rather than the Christian practice of utilizing BC and AD to portray transitions from ancient to contemporary civilizations. Schlafly intended Conservapedia to reflect what he termed the Christian and American value systems. By 2010, Conservapedia had accumulated only 32,000 pages of content, and it clearly posed no serious competition to Wikipedia and its related programs.

One additional new online source of reference materials merits notice, for it also could not have existed prior to the emergence of ICTs. The Electric Literature Directory (ELD) offers a compendium of works, each of which has been created digitally. The ELD archives electronic literature, including prose, poetry, and multi-media games, and it is frequently referred to as *hypertext fiction*. In the 1990s, a number of authors realized they could create materials in hypertext formats on computers or on the Internet. Such a format allows readers to proceed through a text in a nonlinear manner by following links to various sections of the work. Hypertext fiction was promoted as a new manner of creating and reading text afforded by the capability of the computer to permit rapid transfer from one page of text to any other linked page based on the similarity of content.

For example, a romantic novel might begin with a description of the setting or locale and the characters of a story. The reader can switch from the first mention of a major character to another page containing a biography of that person. That page might mention the school attended by that character and include a link to more information about the school or other characters who also attended. Authors were intrigued by the possibilities of new creative forms of literature that could also include interactive audio or video components. Of course, many critics predicted that hypertext fiction would disappear as a passing fancy. However, in 1998, the U.S.

National Endowment for the Humanities provided funds to establish the ELD, and a new and expanded version of the archive went online in 2010. Although the ELD is not affiliated with the Wikimedia Foundation, it does operate on a wiki-like platform. It is also a crowd-sourced archive in that the public is invited to submit entries. However, an editorial board reviews submissions to determine whether they are indeed genuine electronic literature creations.

This discussion makes clear that ICTs facilitate the creation of new kinds of knowledge resources, most noticeably large-scale databases, online reference materials, and a wide variety of data archives. All these new resources expand the kinds of research activities that scholars are pursuing. There is hardly any field of scholarly inquiry that is not affected in a major way by these new knowledge-creation capabilities.

KNOWLEDGE DISSEMINATION

ICTs create many fundamental changes in the patterns of dissemination of knowledge. All publications of books, journals, magazines, and newspapers are currently undergoing major transformations. The old economic models of dissemination are increasingly becoming less viable. Most publishers are struggling to find new and innovative models of operation that will encompass both the older printed formats and the emerging digital and online modes. It is difficult at this point to see just how the dissemination of knowledge will occur in the future. However, it is possible to distinguish two different, yet interrelated, issues that will influence the coming transitions to the new and integrated modes of knowledge dissemination: the transformation of the book and the growth of online publishing. These topics have been introduced in earlier chapters, but they will be revisited in this section from the perspective of the dissemination of knowledge.

Transformation of the Book

The wide-scale production, distribution, and use of printed books have been credited with the creation of all that is encompassed in the notion of "Western Civilization." Without books and the attendant literacy, the Industrial Revolution could never have occurred. The factories that created and then supported the revolution's expansion required a workforce that was both textually and numerically literate, and those requirements eventually gave rise to public school systems that prepared future workers. Books were an essential part of that education. However, today the proliferation of ICTs promotes serious rethinking about both the format and the function of the printed book. As noted in Chapter 2 in the discussion of social change, three major historical developments preceded this current reformulation of the book. The first was the transition from storing writing on scrolls to the pages of a *codex*, a term used to describe the books of antiquity and the Middle Ages. The codex, consisting of pages bound in a hard cover, was a vast improvement over scrolls. Bound books were easier to read, store, and transport. More will be said about the transition from scrolls to books in the discussion of the history of libraries below.

The second major historical development began in the year 1440 when Johannes Gutenberg introduced the printing press with movable type. Prior to Gutenberg, books were written and copied by hand, a slow, tedious, and labor-intensive effort. Printing presses enabled a much more efficient means of producing books, and publishing gradually became a major industry. Booksellers and bookstores proliferated in all the large cities in Europe and later in North America. Well into the twentieth century, moveable type was used in the printing process. Typesetters manually arranged pieces of metal shaped as letters, numbers, and special characters and placed them in a form holder corresponding to each line of text on a page. Eventually, type-setters were replaced by so called "hot type machines." Operators sat at a keyboard similar to those used today with personal computers and typed the text. The hot type machines used molten lead and created each line and page of text. Production speed and efficiency were greatly enhanced by the hot type machines, which became known as *linotypes*. In the 1940s, the linotypes were replaced by machines that produced photoengraved plates for printing. These more efficient machines were also used in the printing of newspapers and magazines.

Currently, the third major transition involves the digitization of printed materials. The first use of digitization in the book industry began in the 1980s when publishers started using computers to drive the printing presses. Rather quickly, the hot type and photoengraving machines became obsolete. A by-product of this new method of printing was a digital file of the contents of the book. When it became possible to transfer rapidly the digital files of books from a central server to a desktop computer or a mobile device, those files took on a new and highly lucrative function, now referred to as the digitization of books. Book publishers and distributors alike saw that it was possible to create mobile reading machines and thus open new markets. Digital copies of books could be sold for a fraction of the cost of hardcover copies or even paperbacks. The high-speed transfer of digital files from publisher to consumer has drastically changed the economics of knowledge dissemination, and the effects are cascading beyond book publishing to newspapers, scholarly journals, and magazines. In fact, the transfer of digital files is affecting all mass media, including film, television, and radio. Any media content that can be digitized can be widely distributed at minimal costs with ICTs, and the volume of such distribution is predicted to increase substantially in the near future.

The Google Book Project has become a major player in the dissemination of knowledge found in books. But first, a few words about the Google company will provide a context for a discussion of the Book Project. As mentioned in Chapter 4, Google is an international company that now dominates the global information markets. It began as an Internet service with a unique and efficient algorithm for locating Web pages with specific search terms. The Google algorithm for Web searching was so superior that it quickly dominated the field. Search engines are computer programs that scan Web pages looking for words or phrases that match the characters requested in a search input. As the World Wide Web expanded, the simple strategy of most other search algorithms was overwhelmed by the growing volume of pages. Google's strategy analyzed the relationships between websites and ranked pages by their importance based on the frequency of citation. Although the specifics of the ranking algorithm are kept secret, Google searches now routinely scan millions of pages in a matter of seconds. In no time at all, people searching the Web became steadfast users of this free service. The word *Google* became a part of common vocabulary to denote Web searching. Children can frequently be heard advising their parents to "Google it" to find the answers to their queries.

Google was founded in 1998 by Larry Page and Sergey Brin, two computer science graduate students at Stanford University. Six years later, it went public, and its stock was listed on the NASDAQ. Google's goal is "to organize the world's information and make it universally accessible and useful." By 2010, the company employed more than 20,000 people operating a public cloud facility consisting of more than 1 million servers distributed all over the world. In terms of rapid corporate growth, Google may be a record holder. The Google search engine processes more than 1 billion daily requests. Approximately 90 percent of Google's $25 billion in annual revenue comes from the advertising of products and services on its Web pages. In addition to the search engine, Google offers an email service (gmail), social networking (YouTube), Web browser (Chrome), photograph manager (Picasa), instant messaging (Google Talk), a mobile telephone service (Android), and now eyeglasses that project a wide variety of images (Google Glass). However, the search engine remains the core of Google's business.

In pursuit of its goal to provide free and readily available public access to all the knowledge in the world, Google launched the Book Project in 2004, the company's most ambitious and controversial effort. The controversy surrounding the project derives from various definitions of intellectual property ownership. This controversy is complicated by the fact that different nations have widely divergent laws and practices with respect to the protection of copyright. As mentioned in Chapter 4, the original intent of the first copyright laws in seventeenth-century England was to encourage authors and artists to be more productive by promising that any revenues that might derive from the sale or use of their products would accrue to them. In recent years, in an attempt to clarify further the protection of the financial interests of the creators of any work, the concept of *fair use* has been introduced. Fair use specifies that a person other than the creator can have limited use of the content for educational or scholarly research purposes. The use of ICTs in the production, distribution, and use of such creations has vastly complicated both the theoretical rationale and the practical application of fair use.

Current copyright law in the U.S. illustrates both the clarity and confusion created by the proliferation of ICTs, and more specifically by the Google Book Project. The implications of this project for the dissemination of knowledge in the future are profound. The current U.S. law specifies that any original work can be protected with a simple copyright claim attached to the newly created object, which could be a book, a document, an article, an essay, a musical score, an image, a CD, or any other form of media. When the work is identified with the copyright sign, ©, it is then protected for a period of 75 years after the death of the creator. Protection in this case means that no one can copy that work for the purpose of his or her personal economic gain without the explicit permission of the creator. The permission to copy may be transferred by the creator to another person or organization, for example, an heir, a publisher, or a recording company. It is no longer necessary to file a copyright claim with a government agency to establish such protection. Simply attaching the copyright sign to the item guarantees protection.

The copyright law in the U.S. has undergone many revisions over the past century, and it seems reasonable to assume that further modifications will occur in the future. In part, some of the need to refine the concept of an original work will force further modifications of copyright law. For example, it is now possible using software to modify photographs to rearrange substantially the content of an image. By what criteria will one be able to distinguish between the violation of the copyright of the originator of the photograph and the creation of a new piece of graphic art? The increasing use of ICTs to create and widely distribute multi-media works of art combining images and recordings of sound and video will surely require new modifications of our conceptions of intellectual property rights and the relevant copyright laws and practices.

Now back to the discussion of the Google Book Project. As mentioned in Chapter 4, Google intends to create an online file that will eventually contain the full text of all books ever published in any language. Furthermore, access to this file will be free and universal to anyone who has a connection to the Internet. To many people, this objective seems overly ambitious. ICTs certainly afford the technical capacity for such an undertaking, but the volume of scanning and digitization is daunting. Nevertheless, by enlisting the cooperation of many of the largest research libraries in the world as collaborators in this effort, Google is making substantial progress in scanning books and creating this enormous file. At the start of the project, Google had enlisted the participation of five of the largest research libraries in the world: the New York Public Library, Harvard University, Stanford University, the University of Michigan, and the Oxford University Library. Since then, 26 other research libraries from the U.S., Europe, India, and Japan have joined the project.

Participation in the Google Book Project means that each library has promised to scan the books in its collections and deposit digital copies in the project files. In theory, this means that all the collections of the world's greatest libraries will be immediately available online at no charge to anyone around the globe. However, in practice, only the full text of those books that are no longer under copyright protection will be available. Bibliographical or descriptive information will be available for books still protected. In some cases, when the publisher or author agrees, a small portion of the book will be available. Nevertheless, the scope of the Google Book Project is unprecedented in terms of knowledge dissemination. In concept, the contents of the Library of Alexandria will be available online. This transformation of the book from a printed format to a digitized, online version has profound implications in many dimensions.

In practice, however, it will be a long time before all of the books ever published can be included in the Google Book Project. Many issues need to be resolved as the project moves forward, but three deserve mention here. Unquestionably, the financial resources needed to implement such a huge undertaking are daunting. Google intends to gain revenue from the project in the same manner it does with its search engine, by soliciting advertising fees. However, additional funds will be needed to support the substantial costs of the project, for example, the scanning operations in all the participating libraries. Private foundations and government funding support the initial stages of these activities. The rate at which books are added to the Google file will undoubtedly be influenced by the ongoing availability of the substantial funds necessary to support the project.

A second major problem for the Google Book Project involves the question of copyright protection as it is applied in different nations. Despite repeated attempts to arrive at universal copyright standards that might apply to all nations, important differences of opinion and practice still remain. It is unlikely that accord will be attained in the near future. Particularly troublesome is the question of how to deal with books that are referred to as *orphans*. As mentioned in Chapter 4, these are books that still have copyright protection, but the owner of that protection can neither be identified nor located. Google claims that all the scanned books it intends to include in its file will no longer be under copyright protection. However, in many instances, this will not be easily determined. Clearly, books that are in print, under copyright protection, and actively promoted for sale will not have full-text copies in the file. The processes of resolving the legal status of orphan books will also add substantially to the costs of the project. Efforts are already underway to determine the copyright status of the more than 13 million volumes in the library at the University of Michigan.

The third concern that has been raised about the Google Book Project is the threat of a monopoly over access to books in the future, which will affect how libraries function. The concern is real, for libraries are shifting their traditional focus from local collection development to providing access to remote materials. This trend began with the growth of the interlibrary loan system. As the costs of books and periodicals have escalated in recent years, more libraries have been relying on loans from other libraries to meet the needs of their users. If the Google Book Project achieves any degree of success, the need for collection development in all libraries will be greatly diminished.

The concern is that access to books will fall under the control of one organization, a publicly owned for-profit corporation. Many corporations during the post-industrial era have engaged in dramatic redesigns to maintain financial viability. If Google should in the future decide to charge for access to the book file, the effect on the world of research and scholarship could be profound and obviously negative. For example, the rate increases of telephone and cable TV companies demonstrate that once charges are initiated for services, they have a tendency to escalate to cover ever-increasing costs. A monopoly on access to knowledge would dramatically change the economic structure of society. As Bell pointed out, knowledge is a keystone of the post-industrial society. Therefore, any restriction of access to knowledge based on a profit motive could constitute a major threat to social order. Important ethical and practical issues need to be addressed. Furthermore, they need to be addressed from an international perspective. Since the reach of ICTs is global, the resolution of access to knowledge cannot be effectively resolved by any one nation. Assuming these problems of funding, copyright, and monopoly can be adequately resolved, one vast collection of online knowledge could become a reality.

Growth of Online Publishing

The phrase *online publishing* includes a number of quite different activities; it is useful to distinguish among three categories: self, combined, and electronic. Furthermore, there are several forms of self-publishing. Many authors post their manuscripts on the Internet and notify all the people who may be interested in its availability. Such documents may be read online or downloaded and printed. Very few of these books receive much attention in the form of promotions or reviews. No one has a reliable estimate of the number of books that are published in this manner, but it seems to be rather small. A much larger number of books are self-published online with assistance from commercial firms that offer a variety of activities to enhance distribution, such as editorial services, indexing, graphical design, page layouts, promotions, sales, and accounting. The authors pay for each of these services. In return, the authors receive a royalty on sales, and these payments are frequently substantially greater than the 5 to 20 percent that hard copy book publishers traditionally offer. This is quite understandable, for online publishers incur few expenses. There are no printing, binding, shipping, promotion, or storage operations involved.

Second, an increasingly common form of publication includes a combination of online publication and hard copy printed books. A number of publishing houses are experimenting with this joint approach to sales and distribution. Preliminary reports indicate that for certain kinds of books, particularly higher education textbooks, access to portions of online copies promotes increased sales of the printed cloth and paperbound books. An option that may prove to be viable is for publishers to grant access to certain sections of texts and limit other parts to printed copy distribution. When textbooks consist of the materials typically covered in a course, previews or summaries of chapters are sometimes available online to help students understand and master the content more effectively. Increasingly, college and university textbooks contain CDs with audiovisual materials to enhance further the learning experience. Laboratory experiments, simulations, expert interviews, and data files are some of the materials included in the textbook CDs. As the capabilities for downloading extensive interactive materials from cloud storage systems increase and the cost decreases, the range and variety of such supplementary text materials will expand. The possibility of combining printed and downloaded content of publications offers new venues for creative expression that can include both interactive and updated components. Another development is the merger of downloadable e-textbooks with the systems that facilitate classroom management tasks, such as student record keeping, attendance tallies, laboratory assignments, grades, and the results of quizzes and examinations. Experimental mergers of publishers and software management systems are being explored.

Third and, thus far, a less frequent pattern of online publishing is the publishing house which previously printed books and has converted exclusively to online publishing. In essence, this happened with many of the encyclopedia publishers who have abandoned printing in favor of online access. In 2006 Rice University Press discontinued publishing printed books and announced that it would continue as the university-affiliated publisher, but all future books would be distributed online. However, in 2010 Rice abandoned the experiment and closed the press. Many other university presses are considering moving in the same direction, perhaps combining printed and online products. All publishing houses, whether commercial or university, are hard pressed to find new means of operating with financial stability. However, university presses, which have always operated on slim margins of income over expenses, are struggling to find ways of sustaining themselves in an economy in which costs continue to escalate. Drawing on the phenomenal financial success of Google with its ever-growing revenue stream from advertising spots on its search pages, some university and commercial presses are considering online publishing supported by advertising. Imagine reading online a book published by a university press in which advertisements for automobiles, deodorants, and cruise lines appear on the pages. It seems clear that the publishing industries and consequently the distribution of knowledge will undergo further dramatic transformations.

Data Curation

Before turning to a discussion of the changing patterns of knowledge use, it is necessary to comment briefly on how the techniques of storing and retrieving knowledge are also influenced by the proliferation of ICTs. An inquiry into the changing definition of the concept of *curate* is useful here. The word comes from the Latin root, *curare,* which means "to care for or oversee." In the Middle Ages, a curate was a person who looked after or cared for the souls of the parishioners. The curate was frequently a member of the clergy who assisted a priest or a rector. In more recent times, the word *curator* has used been used to describe a person who looks after or manages a collection of objects in a museum, an art gallery, or a library. Synonyms include *conservator, custodian,* and *steward.* More recently, the phrase *data curation* has been coined to describe the processes by which digital files are cared for, including their storage, protection, and retrieval to ensure availability for future use.

A data curator then is a person who has the technical skills and experience to deal with such issues as data deterioration and compatibility of past, present, and future hardware and software formats. Many people have had the experience of trying to convert data files stored on an older device such as magnetic tape or a floppy disk to a newer medium such as a CD or a memory stick. Given the speed with which changes in ICTs have occurred in the past, data curation is becoming a new and demanding set of job skills. For example, the task of retrieving subsets of data from the ever-expanding files of the large, longitudinal projects will become more complex each time new files are added. The issue of data protection becomes more complicated, and yet more necessary, as cloud computing continues to expand. In the case of data curation, ICTs have spawned a new and highly technical profession.

KNOWLEDGE USE

Any discussion of the multiple and varied uses of knowledge must limit its scope in order to be comprehensible. This section focuses primarily on libraries, the organizations that facilitate the use of knowledge and that are most dramatically affected by the widespread use of ICTs. However, there is some evidence of ICT-enabled changes in organizations other than libraries. This section also includes brief discussions of changes in museums, and by implication galleries, and archives. Indeed, any organization that collects knowledge in digitized formats has the potential of becoming a virtual research institute supporting scholarly research.

History of Libraries

The gathering of inscribed clay tablets dates back to the some of the earliest known human societies along the banks of the Tigris and Euphrates Rivers in Mesopotamia. One of the earliest recorded civilizations was that of the Sumerians. Lionel Casson, a recognized historian of the ancient Near East, speculates that the first writing on tablets recorded the Sumerians' financial records, including the transfers of goods, monies, lands, and slaves. Gradually the writing of text began to appear on the tablets as records of the names of locations, various animals, occupations, and eventually myths. The archaeologists who discover the remains of such collections use carbon dating techniques to establish that some of them are actually more than 5,000 years old. What is of interest from the perspective of modern-day libraries is that the Sumerians not only collected the tablets in one location, but they also devised a cataloging system for identification and alerts if an item was removed and not returned, an early form of circulation control.

About the same time, writing also began to emerge not far away in ancient Egypt. The Egyptians used papyrus rather than clay tablets as the medium for their writing and record keeping. Made from the dried and pressed core of the papyrus plant that abounded all along the banks of the great River Nile, this early form of paper offered distinct advantages over clay tablets. When papyrus was rolled around a dowel, it was much easier to carry and store. The Egyptians were so successful in growing the papyrus plants that they became the suppliers of the dried and pressed paper to many Mediterranean societies, including the ancient Greek city-states. However, as papyrus was a plant, it deteriorated more quickly than the clay tablets. In fact, during the many wars among the inhabitants of the ancient Near East when the palaces of the vanquished were usually destroyed by fire, many of the tablets were preserved. Archeologists today frequently have more clay tablets than papyrus scrolls to examine.

From the perspective of knowledge creation, dissemination, and use, perhaps the greatest legacy of the ancient Egyptian and Greek civilizations was the expansion of writing and the attendant establishment of the Library of Alexandria, which attempted the grandiose goal of collecting all the writings in the then-known world and making them available to the public. As noted earlier, Bush stated this same aspiration in his description of the Memex Project, and it is also related to the current goals of Wales in his various Wiki activities, as well as to the efforts of Brin and Page in the Google Book Project.

Libraries have been a major institution in Western Civilization, promoting the collection and use of knowledge. Historians generally recognize that the Library of Alexandria was the first large, comprehensive library. Following the death of Alexander the Great in 323 BCE, the huge Macedonia Empire he created was divided into three segments; the largest and most powerful was the land of Egypt subsequently ruled by the Ptolemy Dynasty. Ptolemy I established the Library of Alexandria with the intention of collecting all the scrolls in the world. He was particularly interested in gathering as much of the literature and knowledge of ancient Greece as possible. He sent his agents out to purchase scrolls at fairs in Greek city-states, and there are accounts of occasional thievery. Ptolemy II and Ptolemy III, son and grandson, continued the vigorous collecting. Accurate numbers of the size and scope of the collections are not available, but estimates range anywhere from one-half to 1 million papyrus scrolls. There is evidence that the custodians of the library developed rather sophisticated cataloging and circulation systems. In addition to being recognized as the first large-scale library in Western Civilization, the Library of Alexandria also contributed to the dissemination of ancient Greek as the first language to be written with letters of an alphabet, rather than the hieroglyphic symbols of ancient Egyptian. Of the volumes collected in the library under the auspices of the Ptolemaic rulers, some were recorded in Egyptian hieroglyphics and others in Greek.

The famous Rosetta stone unearthed by Napoleon's soldiers in their Egyptian Campaign of 1799 contained a priestly decree from the Ptolemaic era written in hieroglyphics, ancient Greek, and a cursive language called demotic. The stone permitted the translation of hieroglyphics and opened the door to the study of ancient Egyptian culture and history. From the perspective of this discussion, the Rosetta stone established the co-occurrence of both Greek and Egyptian writing protocols several centuries BCE.

Varying accounts of the destruction of the Library of Alexandria include one where Julius Caesar destroyed it by fire in 48 BCE. However, the library no longer existed by the first century CE. One theory posits that the scrolls were buried somewhere nearby and await their rediscovery by modern-day archaeologists. Perhaps there is a plot here for a new reality TV series.

During the Hellenistic period from the death of Alexander to roughly the first century BCE, literacy and the production of scrolls expanded substantially. Correspondingly, libraries grew in Athens, Rhodes, and many other cities. A school was often connected to the library. Initially, these schools focused on military and athletic training; eventually they became educational facilities that emphasized general teaching and learning. The connection between libraries and education has persisted through time to the present day. As the Romans began to expand their influence in the last centuries of the BCE era, they adopted a version of the Greek alphabet to create Latin, which became the lingua franca of the emerging Roman Empire. The city of Rome became a fertile environment for the creation of a large number of public libraries, the most well-known included the Palatine, Trajan, Porticus of Actavia, and the Library at the Baths of Caracalla. By 350 CE, at least 29 large public libraries existed in the city, and many others were distributed throughout both the western and eastern reaches of the Roman Empire. Clearly, the proliferation of writing on papyrus scrolls spread literacy and knowledge throughout the ancient world. The knowledge accumulated by previous generations was now readily passed on to future students and scholars through the institutions of libraries, thus forming one of the basic foundations of civilization.

The next step in the evolution of libraries and knowledge accumulation was the creation of the codex. Prior to the CE era, almost all writing was done on papyrus scrolls. The Romans first created the tablet notebook to replace the scrolls. The word *codex* in Latin means "block of wood," and writing was done on both sides of the block. Holes were driven in the blocks, and a cord held them together to form a tablet. In some cases, the wood was replaced with a block of ivory. Eventually, the blocks gave way to sheets of papyrus and then parchment, which was more durable and easier to manipulate. Once the cords were replaced with a binding that held the pages together, the book as it is recognized today emerged. The creation of the book enabled a mobility that had never been achieved with the awkward scrolls. The replacement of the scrolls with books dramatically changed the design and operation of libraries.

It took a long time for the bound book with writing on both sides of the pages to replace the scroll as a medium for storing and using knowledge. In fact, nearly 500 years had passed before the scroll disappeared from common use. No one knows for certain why this transition took so long, but the record is clear that from the earliest centuries in the CE era, all religious texts produced by Christian writers were books rather than scrolls. Some evidence suggests that although most, if not all, religious or spiritual texts were recorded in books, whereas financial and legal documents continued to be recorded on scrolls. However, by 500 CE, the scroll had all but disappeared.

When the Roman Empire collapsed in the fifth century CE, it was divided into two portions, the Eastern or Byzantine and the Western or Christian. Particularly in the Christian region, libraries were associated with monasteries, institutions that flourished into the Middle Ages. A number of the founders of these monasteries, including Pachomius, Benedict, and Cassiodorus, declared the monks must be literate, for it would allow them to read, understand, and proselytize. Therefore, instruction in "letters" was deemed mandatory in the daily routine of the monastery, and it involved the monks as either students or teachers. The emphasis on literacy as an important aspect of the life of the spirit entailed creating libraries and the scriptoria, rooms where the monks copied religious texts. Scholars of the period leading up to the Middle Ages credit these monasteries with translating into Latin the texts of the ancient Greeks and Romans, thereby preserving these writings for later generations. Thus the works of Homer, Aristotle, Plato, Hippocrates, Galen, and many others remain available today.

The monasteries that spread throughout Byzantine and Christian regions solidly created the juxtaposition of libraries and teaching and learning. Particularly in the Christian areas, cathedral schools were established to teach literacy to the local population in support of both the sacred and secular functions of the congregation. When the first universities were created in Italy and France in the eleventh and twelfth centuries, libraries were a natural adjunct to the academic programs. The pattern of libraries associated with educational institutions at all levels persists to this day. During most of the Middle Ages, the collections of libraries in cathedral schools and subsequently in universities consisted of books that had been produced or copied manually. As discussed earlier, the invention of the movable type press vastly increased the production of books and thereby expanded the economics of the book trade business, as well as the amount of materials collectible by libraries. By the late Middle Ages, the profession of the librarian began to emerge with specific tasks supporting acquisition, circulation, reference services, and preservation.

This digression into the history of literacy and libraries provides a background to demonstrate the escalating rate of change in the creation, dissemination, and use of knowledge. Approximately 5,000 years ago, the Sumerians kept financial records on clay tablets, and the Egyptians recorded hieroglyphics on papyrus scrolls. Approximately 2,000 years ago, libraries in Alexandria, Egypt, and many Greek and Roman cities collected scrolls written with alphabets. At about the same time, the book, first called a codex, began replacing the scrolls. Approximately 1,000 years ago, monks in monasteries produced books. Approximately 500 years ago, printing presses began large-scale production of books. Approximately 50 years ago, the computer was invented, and the age of digital information was launched. Approximately 10 years ago, the Internet was invented, and the processes of knowledge creation, dissemination, and use began to be shared around the globe. Surely this escalated rate of change of the knowledge processes will continue in the future, and the structure and function of knowledge institutions such as libraries will continue to adapt to the new information environments. Certain functions of all libraries such as acquisition, cataloging, circulation, and preservation were evident in the libraries of ancient times, and they have persisted in altered forms down through the ages.

Academic Libraries Present and Future

A major shift is occurring today in the focus of academic libraries. Not too many years ago, these libraries devoted a substantial portion of their resources to purchasing materials necessary to support the research and instructional activities of their local faculty and students. Many university libraries had large staffs of full-time employees who devoted their time to identifying and acquiring the books, journals, and other materials that they and faculty members deemed important. The library staff, usually referred to as *bibliographers*, possessed expert knowledge in the subject areas of interest to the local faculty. Many bibliographers were motivated to accumulate the most comprehensive collection of all the relevant literature in their fields, past, present, and future. For example, one land-grant university prided itself on collecting all the research documents ever published on the growth of mushrooms in all U.S. states east of the Mississippi River.

In recent decades, the cost of acquiring materials, particularly academic journals, has escalated; library directors and university administrators became alarmed at the ever-expanding costs of collection development. Some years ago, the library director at a large research university studied patterns of use of the library's holdings and discovered that 80 percent of items purchased by the library had never been circulated five years after they were added to the collection. Librarians, as well as university financial officers and presidents, had to question whether they could afford to continue such acquisition policies. As the financial resources available to both public and private universities have dwindled in recent years, the answer to that question was resoundingly negative. The interlibrary loan system was one attempt to contain these costs. However, it was never widely used, because too many faculty members wanted the materials close at hand, and they were unwilling to wait for something to be delivered on loan from another library. Given the increasing availability of online materials and the growing constraints on budgets, interlibrary loan rates have declined in recent years.

The focus of most academic libraries is shifting from acquiring materials to add to the local collection to providing access to materials that are increasingly becoming available remotely in online digital collections. The growing expansion of projects such as the Google Book Project, the repositories of digitized academic journals, and crowd-sourced online reference files such as Wikipedia will eventually make available to scholars and students alike large portions of the contents of the academic libraries. The format of the materials used by academic libraries now includes a wide variety of multi-media materials. In part, this shift reflects the soaring costs of printed materials, especially scholarly journals. However, it also reflects the fact that ICTs enable the production of multi-media materials that faculty members need for research and teaching. Unfortunately, the costs of these newer materials are also high when the expense of all the hardware and software necessary to store, maintain, and retrieve them is included.

As mentioned earlier, the Google Book project to create one database of all printed materials is both ambitious and controversial. A major issue concerns whether access to that database in the future will continue to be free to the user. A time-honored tradition in the profession of academic librarianship is that users pay no fees for access to the collection. Will that philosophy continue to pertain to such large-scale repositories as the Google Book project? The resolution of that question will have a major impact on the future of academic libraries and scholarly research.

The Association of College and Research Libraries (ACRL) released a report in 2010 reviewing trends in academic libraries, and it called attention to the constrained economic resources available generally to institutions of higher education and specifically to academic libraries. It acknowledged the shift in the collection development policy from gathering materials "just-in-case" they are needed to providing remote access "just-in-time" as the need arises. The report also discussed the increasing digitization of library collections, such as the special collections of unique materials relating to the American Civil War under the auspices of the Association of Southeastern Research Libraries, or the California Local History Digital Resources Project involving 65 institutions working under the direction of the University of California Digital Library. The report discussed the growing use of mobile devices among students and faculty and the expected demand for reference services using handheld devices. The report anticipated new roles for academic librarians as collaborations and communications both within and among institutions of higher education increased. All these interactions are facilitated by the social networking afforded by ICTs. Finally, the report discussed the changes occurring in the use of physical space in academic libraries. As less frequently used materials are stored in remote locations, it is anticipated that those spaces formerly used for materials storage will be refurbished into areas for expanded services, such as media centers and tutorial or remedial activities.

In recent years, several academic librarians have promoted what they call *blended librarianship*. The concept implies that blending the perspectives, expertise, and skills of instructional design, technology, and traditional librarianship will open new avenues of practice and professional development for academic librarians. As university libraries move away from local collection development to providing remote access to materials to support research and instruction, the traditional roles of the bibliographer, cataloger, and reference specialists are undergoing transformations. Blended librarians are new professionals who are knowledgeable in the traditional roles, but they also develop competence in information technologies and curricular design. For example, they might address the issues of instructing students in the importance of evaluating the accuracy of information, rather than simply taking what they happen to find on the Internet as correct and valid. Blended librarians would help students become more critical and sophisticated users of the vast troves of data in cyberspace.

The blended librarian might also work collaboratively with faculty to use more effectively both the content resources relevant to their courses as well as the technology that can be integrated into the curriculum and the students' learning experiences. Some libraries assign staff members to departments to work more closely with faculty and students. This pattern of embedded librarians working in academic departments is emerging initially in the sciences, but it is expected to occur in the social sciences and humanities before too long. Clearly, the blended librarian is an attempt to fill a gap in the use of knowledge in college and university environments. The word *environments* is used purposely here instead of campuses, for an increasing amount of higher education teaching and learning occurs online and not in the traditional ivy-covered buildings.

Several academic libraries, most notably that at The University of Chicago, are resisting the option of remote storage of large portions of the less frequently used items in their collections. Chicago has recently opened a new building that uses technology to house and retrieve its 3.5 million volumes. It is adjacent to the main library building, and they are connected by a walkway. The new building stores books by their size, rather than by the Library of Congress call numbers, which is the system used by most academic libraries. The volumes are stored in 24,000 bins in 10 columns of racks that extend 50 feet below street level. Temperature in the underground storage area is maintained at 60 degrees to preserve the holdings. Five commercial cranes are used to move between the 10 columns to access the bins. Library users request volumes online, and the cranes move to the appropriate bins and deliver the items to a circulation desk. The user gets an email notification that the item is available, and the entire retrieval process takes less than five minutes. The cranes redeposit the volume when it is returned to the library. The new storage facility cost $81 million, and the academic library community seems divided on whether the cost justifies having less frequently used materials available with a five-minute recall.

A final note is called for in discussing a wider impact of ICTs on knowledge use in organizations other than libraries. Increasingly, museums of all types are rethinking their missions and expanding their programs to incorporate the capacities of ICTs to enable interactive exhibits that allow visitors to experience their collections in cyberspace. The New York Metropolitan Museum of Art recently reported that more people visit remotely than physically come through the building's doors. The education programs of this and many other museums are expanding to take advantage of ICT-enabled technologies. Google recently announced its Art Project, which uses the Street View software that is employed in Google Maps. It allows users to see a real-time picture of almost any street address on the globe. In Art Project, the software allows users to view major pieces of art in 17 of the world's largest art museums, often accompanied by explanatory text files and YouTube videos. With the use of special cameras, the projected art works have amazing clarity, allowing the viewer to distinguish even brush strokes. The Google Art Project significantly expands all the participating museums' capacities to reach out and engage larger numbers of geographically dispersed patrons. Similar distant-viewing capabilities will appear soon in commercial art galleries and archives that house documents, art works, and other ephemera.

The virtual discipline institute is another type of knowledge distribution organization that is appearing more frequently. Scientists now use the Internet to conduct investigations and share results with colleagues around the globe. The speed with which announcements of new discoveries in physics travel among colleagues and then get reported in the public media has recently caused some problems requiring retractions to set the record straight. Professional associations now create online access to resources that make knowledge available to larger audiences. For example, the American Sociological Association has recently made available to its members and the public an online collection of syllabi for all the courses that are typically offered in colleges and universities.

This chapter has documented many new patterns of activities in the creation, dissemination, and use of knowledge that are afforded by ICTs. Some older organizations have modified their operations, new scientific organizations have been created, new businesses have appeared, new professions are emerging, and many scientists and scholars are changing the topics they study and the methods they employ in their work. Without ICTs, all these changes and new developments would be unattainable.

7   Consumerism

The economies of developed nations depend heavily on an ever-expanding consumer society. The purchase of goods and services is the foundation of a healthy modern economy, and advertising is the driving force of consumption. According to the U.S. Census Bureau, national retail sales more than doubled between 1992 and 2010. Advertising, which is designed to increase mass consumption, has also grown steadily. However, it is the proliferation of information communication technologies (ICTs) that enables the growth of online advertising. In 2011, expenditures for online advertising were expected to exceed $32 billion, and the industry doubles that amount every five years. Furthermore, the amount of money spent on online advertising now exceeds that spent on the major television networks. The growth of social networking, mobile computing, and television viewing on the Internet promotes advertising that is increasingly targeted to individuals' patterns of past online behaviors. These developments have given rise to what is now referred to as *eConsumerism*.

This chapter examines the current explosion of consumerism as facilitated by ICTs. First, it briefly explores the history of mass production and consumption as a consequence of the standardized manufacturing techniques that grew out of the Industrial Revolution. It then examines the role of the advertising industry in promoting the consumer society. The next topic discussed includes the new means of buying, selling, and bartering enabled by ICTs. The chapter, then, describes a number of examples of the new products and services whose consumption is possible only with ICTs. Finally, the chapter briefly explores the controversial concept of the post-consumer economy that has been proposed as a plan for future development.

MASS PRODUCTION AND CONSUMPTION

Patterns of mass production and consumption emerged gradually as the Industrial Revolution expanded in Europe and North America. One of the hallmarks of mass production was standardization of the shape and size of machine-produced parts. In the U.S. the gun-manufacturing industry was one of the first to adopt standardization. Until the early nineteenth century, gun production was a highly individualized process. Each rifle or pistol was unique in design and size of its component parts: the barrel, the stock, the ignition mechanism, and the trigger. The ammunition was also unique, for it had to both fit into and rapidly exit the bore of the barrel. Consequently, gun and ammunition making were handcrafted processes usually done by blacksmiths. Production was time consuming, inefficient, and costly. Furthermore, the bullets for one gun were not interchangeable with those for other guns. Eli Whitney, the inventor of the cotton gin, was a major figure in the introduction of standardization to gun manufacturing. He used precision machinery to make standard-size units for each rifle part, thus making all parts interchangeable. Whitney perfected the techniques of standardization in his factory in New Haven, Connecticut, laying the foundation for the later creation of the assembly line model of mass production.

Around the beginning of the nineteenth century, Whitney demonstrated his manufacturing techniques for the U.S. Secretary of War and his staff. Placing 10 piles of unassembled rifle parts in front of his audience, Whitney then proceeded to assemble 10 rifles, drawing one part randomly from each pile. The demonstration was so successful that Whitney immediately received a government contract to produce 30,000 rifles for the U.S. Army. Partnering with another Connecticut gun manufacturer, Simeon North, Whitney then went on to help establish two federal arsenals for the mass production of rifles, one in Springfield, Massachusetts, and the other in Harpers Ferry, Virginia. Both arsenals subsequently played major roles as armament producers for later U.S. conflicts in the nineteenth and twentieth centuries.

The standardized manufacturing process spread rapidly from guns to other industries, including those that produced farm equipment, steamboats, locomotives, and eventually automobiles. During the nineteenth and twentieth centuries, population growth and large-scale migration from Europe to North America expanded the pool of potential customers in the U.S. and Canada. Standardization and mass production became the dominant mode of manufacturing and expanded to include household furnishings and a vast array of personal items. The economies of scale gained by producing many items lowered the prices of individual pieces and made them more attractive and affordable for more persons. And to keep the economy growing, sales had to increase. Under these conditions, the birth of the advertising industry was inevitable, and today it is a major component of the economy in all developed nations and a rapidly growing component of the economy in the developing nations as well.

ADVERTISING AND THE CONSUMER SOCIETY

Advertising is truly ubiquitous. Residents of developed nations today are literally bombarded with advertisements encouraging them to buy, lease, or obtain by almost any means a huge number of products or services. Advertisements appear everywhere, in newspapers and magazines, on radio, television, highway billboards, and almost every commercial Web page. Inducements to buy expensive products such as automobiles or furniture with extended time payments promise low interest rates, cash back bonuses, or seller payment of sales taxes. Advertising agencies and the companies that employ them to promote their products and services appear convinced that repeating their messages over and over again in multiple formats increases sales and revenues. Consumers are confronted with a cacophony of speech, text, pictures, videos, music, and jingles. Little effort is made to filter the content of advertisements to eliminate anything that might be inappropriate for children and adolescents. As mentioned in an earlier chapter, Google's major source of income is the revenue it collects from all the advertisements appearing on the pages produced by its services. Online advertising produces revenue for many of the social networking sites as well.

Before turning to the discussion of the new means of buying and selling as enabled by ICTs, it is important to recall the discussions in earlier chapters of the 24 hour/7 day direct access to the Internet. In this chapter, 24/7 implies that the buying and selling activities that used to take place during normal retail business hours can now be accomplished at any time and from any place with Internet access. This vast increase in the total hours in which retailing can take place forms the context within which to examine changes in the patterns of commerce afforded by ICTs.

NEW MEANS OF BUYING AND SELLING

The discussion of the influence of ICTs on the rise of consumerism in the remainder of this chapter draws primarily upon information from North America. However, it is likely that these patterns also exist in other developed nations. This section of the chapter discusses five topics in the category of ICT-enabled means of buying and selling: online retailing, online auctioning and bartering, online classified advertising, mobile computing, and behavioral targeting.

Online Retailing

Amazon, the online retailing empire is the best example of the new means of ICT enabled consumption. This company pioneered online retail selling. Founded in 1994 by Jeff Bezos, with headquarters in Seattle, Washington, Amazon is the world's largest direct online retail merchant. In 2012, Amazon reported more than $39 million in net income and employed more than 88,000 persons worldwide. It began as a bookseller, based on the idea that an online bookstore could offer a much larger number of titles for sale than any chain of stores. Furthermore, the economy of scale enabled by nationwide and eventually worldwide sales would result in both lower prices and greater profits. Amazon has since expanded to become a general merchandise retailer. In addition to books, it now sells a wide variety of products, including computers, electronics, home and garden tools, toys, groceries, clothing, shoes, sports and outdoor equipment, and automotive accessories. It also operates retail websites for other companies, including Target, Timex, Marks & Spencer, Lacoste, and Sears in Canada. The Amazon servers that drive the distribution and warehousing systems are located in North America, Europe, and Asia. Collectively they handle an estimated $1 million in sales each hour.

One of Amazon's recent retail products, the Kindle, is radically changing the way books and other printed media are distributed. Launched in 2007, the Kindle is an electronic book, or e-book reader, which allows users to download a wide variety of digital files into a handheld device for convenient and portable reading. Other companies have also introduced devices capable of reading downloaded content files, including the Reader from Sony, the nook from Barnes & Noble, and the iPad from Apple. At least two dozen additional e-book readers are now on the market with varying download capabilities and display features. The competition among the producers of these and other handheld devices appears to be intensifying and is predicted to grow even more in the near future. In 2010, Amazon reported that 143 e-books were downloaded for every 100 hardcover books the company shipped. The convenience of downloaded sales is increasing overall consumption. Amazon is dramatically changing the way books, newspapers, magazines, and even videos are purchased and consumed. Although Amazon was the pioneer in online selling, almost all other large, and even many small, retail businesses have jumped on the bandwagon, offering their customers the option of online purchases. It goes without saying that such transformations are possible only with ICTs.

Prior to the advent of ICTs, retailers maintained one or more stores for direct face-to-face sales to their customers. Many stores were specialty shops, and others were known as department stores, for they maintained separate departments for a wide variety of retail items. In addition, most large-scale retail businesses mailed printed catalogs to their regular customers. These catalogs depicted their inventories complete with pictures, descriptions, prices, and shipping charges. Customers could visit local stores or place catalog orders by telephone or mail. The annual or semiannual catalogs of companies such as Sears Roebuck, Macy's, Nieman Marcus, and Dillards were highly anticipated events for many families, and the catalogs were often used for planning purchases for special occasions or holiday gifts.

Customers were frequently enticed to purchase items on a layaway plan. By making a small deposit, a customer could ensure that the retailer would hold a particular item until full payment was made. Subsequent partial payments were made on a monthly basis. Retailers organized clubs to help customers make regular payments to a savings account to ensure they had money to spend during the holiday seasons. In this way, the customers could rely on having items available when they had the money to pay in full, and the retailer had more immediate access to their money.

The catalog and mail-order business became a substantial segment of the retail sales industry. More recently, department stores expanded to include retail discounters such as Target, Wal-Mart, and Kmart. Several discount retailers such as Costco and Sam's Club have added extensive arrays of food and grocery items. The latter usually charge some small annual membership fee for access to items in bulk that cost substantially less than traditional retailers charge. The websites of these retailers offer larger arrays of products than the printed catalogs. Using the multi-media capabilities of ICTs, the websites contain greatly expanded pictures and frequently video and audio formats. Customers can place orders instantly on the websites using credit cards. The ease with which these purchases can be made further expands the ever-growing volume of consumerism.

Online Auctioning and Bartering

Auctions and bartering constitute the second category of consumer activities that are enhanced by ICTs. Both types of economic exchanges have existed for thousands of years. As mentioned in Chapter 2, bartering was a widespread practice in farming communities, particularly before the introduction of currencies to facilitate economic exchanges. For example, a farmer might trade two calves for three weeks of fieldwork during the next harvest. Also, auctions commonly took place, especially in rural areas. Auctioneers roamed the countryside, particularly at harvest time, to help farmers sell produce and livestock. Substantial amounts of bartering and auctioneering now take place online, and they are evolving into new and expanded forms.

The largest online auction company is eBay. Founded in 1995 in San Jose, California, by entrepreneur Pierre Omidyar, eBay began as an auction site where customers placed their bids on one or more items before a set closing time. The highest bidder won the auction and paid a small administrative fee. However, eBay expanded its auction options to include reporting the current highest bid during the active period. This policy resulted in a complex system of reporting the second highest bid, which caused a great deal of confusion among both sellers and buyers and a number of lawsuits to resolve conflicts. As a result, eBay added a regular online shopping option where sellers list one or more items and buyers can make an immediate commitment to purchase. If the seller accepts the price, the sale is final. Otherwise, they negotiate for the final price. This eBay option is similar to an online retail transaction, and it vastly increased volume and revenue. Within two years, eBay was facilitating more than 2 million auctions a year.

From 1998 to 2008, Meg Whitman served as president and CEO of eBay, presiding over its global expansion to serve hundreds of millions of users in 30 countries. In its early days, eBay employed only 30 staff, but by 2012 it required more than 28,000 persons to support its operations. Both individuals and businesses now buy and sell a wide variety of items online with eBay. The list of services has expanded beyond the sale of products to include classified advertising, management consulting, and commercial payments using PayPal, an electronic funds transfer system. As with Amazon, eBay adds substantially to consumption in the developed world. However, eBay is only one, albeit the largest, of many online auction sites. Some specialize in certain categories of items, such as antiques, haute couture, or collectibles; many are general auctions in which sellers of all types of products are invited to participate. Some sites charge membership or posting fees, but most others are free. The purveyors of these online auctions gain additional revenue from advertising posted on their websites. For example, gun auction sites might carry the advertising of ammunition manufacturers or hunting clothiers.

A particularly interesting variation of auction sites is 1stdibs.com, read as "first dibs." This auction site specializes in high-end items, including interior decorations, furniture, clothing, rugs, jewelry, watches, and artworks. It differs from eBay in that it serves as the online auctioneer for a large number of retailers that cater to a more affluent clientele; 1stdibs facilitates the listing, bidding, negotiating, payment, and shipping for all the small retailers. The customers of 1stdibs are the retailers, and they make it easier for their customers to engage in visual shopping and purchasing. Thus, the consumer is aided in making purchases from retailers that would otherwise deal only with those who visited their stores in person. Such an arrangement further increases sales volumes.

Bartering or swapping goods or services is another type of business that is growing as a result of expanding ICTs. Although no large companies dominate bartering activities the way Amazon dominates retail sales or as eBay does in auctions, a number of groups hope to expand bartering and change dramatically how consumers behave. It is difficult to estimate the amount of bartering that takes place online, but numerous websites aspire to entice more participants. There is some hope that at a time of sustained unemployment, a surge of interest in bartering as a means of converting property, goods, or services into useable assets will occur.

Online Classified Advertising

The largest online classified advertising company, Craigslist, was started by Internet entrepreneur Craig Newmark as a hobby in 1995. Craigslist began as email exchanges among friends in the San Francisco Bay Area and served as an information-sharing medium about local community events. A year later, it became a website and broadened its scope to include postings on jobs, housing, items for sale, and personal notices. In 1999, Craigslist became a private, for-profit organization that charged varying fees for different postings. Employers seeking to fill job vacancies paid the highest fees. In 2000, Jim Buckmaster, a colorful, charismatic, and controversial former programmer, became the CEO of Craigslist. Under his leadership, the company has expanded well beyond the San Francisco Bay Area to include more than 700 cities in more than 70 countries around the globe. Craigslist claims that it needs only 32 employees to manage this vast global empire of online activity, which processes 20 billion page views per month.

The content of Craigslist postings has also expanded to include discussion forums on a wide variety of topics such as nutrition, health, religion, and politics. The housing category now includes rooms, apartments, houses, and vacation rentals and exchanges. It also includes a section on "gigs" in which people look for someone to do short-term jobs or services such as cleaning, painting, and other household or small business chores. Craigslist maintains its look as a casual, informal site that still uses primarily text rather than elaborate graphics and multi-media formats. Despite the enormous volume of its business, the ambience of Craigslist continues to reflect its origin as a community service. In 2004, eBay purchased a 25 percent interest in Craigslist from a former employee, and legal disputes between the two companies concerning the financial value of their relationship continue.

In recent years, Craigslist has come under increasing criticism for its postings in the category of adult services. The attacks came from a number of groups, including religious organizations and the attorneys general of several states, who claimed that Craigslist promoted prostitution, and that inappropriate materials were readily viewed by children. Adult service postings were estimated to make up 30 percent of the company's revenue. Craigslist responded by placing disclaimer statements on a number of its categories and asking users to state that they were at least 18 years old. Nevertheless, a young woman contacted through a Craigslist posting was savagely murdered, allegedly by a young medical student in a Boston hotel room. The student took his own life in a jail cell while awaiting trial. Both deaths received extensive media coverage, and Craigslist was always mentioned in the news releases. The pressure generated by this case and the increased efforts by several states caused Craigslist to block access to the adult services page on its website. Rather than withdraw the page entirely, Craigslist placed the boldfaced message "CENSORED" in the space previously occupied by the link for adult services. Obviously, Craigslist wants its users to recognize that the public outcry resulted in a form of censorship.

This incident demonstrates that ICTs are responsible for increasing the salience of an issue that has always received the attention of free speech advocates as well as those who are convinced that restraints must be placed on the circulation of salacious materials, especially to children. The legal profession and the court systems have wrestled with this dilemma for many years. The resulting practices at any given time have always been some compromise to placate opposing parties. However, ICTs and especially the Internet have highlighted this issue, for they make it possible for much larger numbers of people to access such materials. It is more difficult, if not impossible, to prevent children from viewing pornography. This is not a new issue, but ICTs have made it more visible and less tractable.

The phenomenal growth of online classified advertising has had a devastating impact on the financial state of the newspaper business. Prior to the advent of ICTs, daily newspapers derived a substantial amount of revenue from classified advertisements. It was a common practice for anyone wanting to buy or sell single products or services to place a small notice in his or her local daily or weekly newspaper. Companies or agencies seeking new employees or job seekers used the classified sections of their local newspapers to advertise. Fees were usually nominal, but in bulk the income to the newspapers was significant. However, people placing advertisements get much greater exposure for little or no fee by using online postings. Newspapers have been hit by a double blow attributable to ICTs. First, their revenues have decreased from this loss of printed classified advertising to online postings. Second, as noted in Chapter 6 on knowledge, circulation and subscription income are falling as more readers, particularly younger ones, increasingly turn to other forms of mass media, including, of course, the Internet.

Mobile Computing

Mobile computing also provides new ways to buy and sell. The proliferation of handheld computing devices and wireless network connections is producing an explosion of continuous access to the Internet regardless of location. Mobile computing enables buying and selling regardless of time or place. Devices equipped with a global positioning system (GPS) provide numerous additional capabilities. GPS is the navigation system developed originally by the U.S. military services to aid in the global positioning of their weapons and units. The system is now available to the general public via GPS receivers, which are widely used in all sorts of locations. It is probably most widely recognized as an aid to automobile drivers traveling to and from any location, but its many potential future applications will enhance additional consumption.

An interesting and relatively new development in computer applications is augmented reality (AR). AR has many applications, but the one that is most pertinent to this discussion of consumerism adds computer-generated information to the view of a handheld camera. Combining mobile social computing with both GPS and AR dramatically expands available activities, thereby promoting further consumerism. For example, GPS receivers in handheld computers can connect to retail stores to find addresses, hours of operation, and items in ongoing sales; determine if an item is available in a specific size or color; place an order; or even compare prices across several stores. Such a system could certainly increase impulse buying among a large number of consumers. Alternately, the system can display the customer rankings and menus of all the nearby restaurants; take reservations; and give walking, driving, or parking information. Or users can find the programs, schedules, and ticket availability of shows, movies, or museums. The opportunities for increasing consumerism with ubiquitous GPS and handheld devices appear to be infinite.

Behavioral Targeting

Behavioral targeting is also known as targeted advertising. The practice of monitoring and storing the online activities of users is now commonplace among both search engines and online retailers. Information such as sites visited, time spent on Web pages, and items purchased with price and method of payment are all routinely collected by Web browsers and online retailers. This information is then analyzed, packaged, and sold to marketers to enable them to target their advertising. The advertising industry is in the process of transitioning from campaigns that are directed at mass audiences to targeted advertisements based on the data collected from past online activities. Targeting is not new; for many years, advertising campaigns ran slightly different copy for various regions of the country. For example, an advertisement for automobiles used different copy in geographical regions that experienced severe or warm weather at different months of the calendar year. Or an advertisement ran different images in a weekly general news magazine from one covering professional sports. What is new is that advertising can now be targeted to individuals or, more accurately, to specific computers.

Almost all search engines and retail websites deposit a cookie on any computer that accesses them. A *cookie* is a piece of text that can record information about the online behavior of the user, and advertisers use that information to target advertising. For example, a user may enter a retailer's website and look at a product, perhaps a household item such as a food processor. The person decides not to purchase the item, but the cookie records the visit and saves the address of that computer. Subsequently, when a user of that same computer visits the Web pages of other retailers, he or she notices advertising for that same food processor. Or when a user visits a website to obtain a weather forecast for a specific locale, a zip code or address is required. That zip code or address is then stored, and the user later finds advertisements for hotels or restaurants near that location. The advertisements might appear on the Web pages of retailers or on the site that the user has selected as his or her home page, which appears every time that computer starts up.

Those who promote the protection of individual privacy are dismayed when they learn of the extent of such data collection and behavioral targeting. Is this an invasion of privacy? Intense debates attempt to resolve this question, and a consensus does not seem likely to emerge in the near future. The behavior of the computer user is not the same as the behavior of a particular individual. However, if one and only one person uses that computer, then the information gathered by the cookie is about that person. Neither the cookie nor the browser that deposited the cookie can know the exact identity of the person using the computer. Therefore, some claim that this does not constitute an invasion of anyone's privacy. On the other hand, many worry there is no guarantee that future systems will not be able to link online behavior to specific individuals by combining the information gathered by the cookie with the data used to process credit card payments. If that becomes a reality, then a company or a combination of companies might have data that could be subpoenaed by a government agency or even used for a variety of illegal purposes. It seems clear that the spread of targeted advertising will promote consideration and eventual revision of the ethical and legal conception of privacy.

NEW PRODUCTS AND SERVICES

ICTs are enabling a wider range of new services and products to strengthen further the foundation of an economic system dependent upon ever-expanding levels of consumption. Hence, consumerism is continually reinforced by the expanding capacities of ICTs. The list of new services and products is long, and the discussion here can address only some examples of the many consumer innovations enabled by ICTs. Indeed, the field is growing daily, and it would be impossible at any given time to enumerate a complete listing.

The Long Tail of Consumerism

A key concept for understanding many of the new activities encompassed in consumerism is that of the *long tail*. The concept was first popularized in 2004 in an article written by Chris Anderson, editor in chief of *Wired* magazine. Two years later, Anderson expanded on the concept in a book entitled *The Long Tail: Why the Future of Business Is Selling Less of More*. The concept draws on the work of the early-twentieth-century Italian economist Vilfredo Pareto. In his studies of the distribution of land and wealth, Pareto noted that typically 80 percent of the land or wealth was owned by 20 percent of the population. Anderson applied this concept to the distribution of retail sales: 80 percent of income is derived from the sale of 20 percent of the inventory. A corollary of this observation is when production and distribution costs are low, substantial income can be obtained from the sale of the other 80 percent of the inventory. Anderson quoted an Amazon employee who described the long tail strategy as gaining substantial revenue by selling those books in the inventory that are not best sellers, rather than focusing on the most popular titles. This is sometimes referred to as marketing to a specific or targeted audience, and ICTs greatly facilitate the implementation of such strategies, particularly when sales are substantial.

When volume is high and distribution costs are low, selling items over the Internet is more profitable than distributing them via retail stores. For example, filling orders for doctor-prescribed medications is moving from traditional drugstores to companies that accept online orders and mail the drugs directly to the consumer. A number of companies now routinely accept doctors' original prescriptions and automatically remind customers when it is time to order refills. These companies also offer at discounted prices a large number of over-the-counter medicines and other health care products previously sold by traditional drug stores. Furthermore, online companies also offer online health care information, including advice on possible interactions among different drugs, nutritional counseling, and exercise routines. In addition, some of these companies offer prescription drug insurance plans. ICTs make it possible for these companies to offer comprehensive health care packages. This transition has been called the "commercialization of health care services and information." As the baby boomer generation matures and the population of the U.S. is aging, health care services to senior citizens are potentially quite lucrative.

Niche Selling

Niche selling in the clothing and accessories industries is another area of retailing that is incorporating ICTs in its advertising and product delivery operations. One of the largest companies doing so is Like.com. Launched in 2006, it offers users the opportunity to examine pictures and compare prices of women's and men's clothing, shoes, and accessories. The Like database contains information on more than 2 million products from 200 merchants. In 2010, Google acquired Like, indicating that the technology giant believes there is a future in such retailing innovations. A number of other retail shopping sites offer visual shopping capabilities, including Polyvore, oSkope, thefind, and Zappos. Other services focus on specific items; for example, Endless offers visual comparative shopping for shoes for women, men, and children. StoreAdore offers an interesting variation on comparative visual shopping. Launched in 2007 by Harvard Business graduate Meredith Barnett, StoreAdore quickly expanded to more than 2,400 boutiques offering customers the opportunity to examine pictures and compare prices on women's clothing and accessories in all the major, exclusive shopping areas in North America. It is tied into both Facebook and Twitter so that customers can interact with friends and seek opinions and advice. These connections to social networking allow customers to simulate group shopping with friends to get their advice or reactions to items they are considering buying.

StoreAdore reports that its member boutiques and customers are delighted with this ICT-enabled enhanced mode of retailing.

The Bigelow Tea Company provides another example of interesting use of ICTs in promoting and distributing products. Founded during the Great Depression of the 1930s, Bigelow is a family-owned business that offers customers a wide variety of teas and related products ranging in price from a few pennies per cup to many exotic and expensive varieties, blends, and accoutrements. Traditionally, Bigelow sold its products through national and regional supermarket chains. The company is currently under the leadership of Cindi Bigelow, granddaughter of the founder, Ruth C. Bigelow. She is leading the company in new directions of growth and collaboration, including expanded distribution, advertising, and customer support and education. Using the social networking capabilities of ICTs, she appears on the company website in videos interacting with a variety of regular and potentially new customers. Bigelow uses YouTube, Facebook, and Twitter to contact her audience. She reports that despite a weak economy, the company is experiencing new growth and now grosses more than $100 million each year.

The Proctor & Gamble Company has taken a different approach with ICTs to promote its products, establishing a website for teenage girls, beinggirl.com. It promotes two of its products, a sanitary napkin and a tampon, which are of interest to girls as they enter the developmental stage of puberty. The site is intended to engage the attention of its visitors and potential customers with brief articles on topics of interest to teenage girls, such as physical body changes, exercise, health, nutrition, music, clothes, dating, and dealing with friends and boys. Visitors are invited to join a Facebook group and participate in a blog. The site also offers advice on how to participate in an online group in a manner that does not put them at risk for stalkers or other dangers. Although beinggirl.com is clearly a scheme for attracting and keeping new customers, it does make a substantial effort to provide sound advice on a wide variety of topics that typically confront teenage girls. Proctor & Gamble is commended by many public advocacy groups for combining its advertising with public education efforts.

On Demand Entertainment

As mentioned in Chapter 4, the music and film industries have undergone radical transformations due to the ICT-enabled capacity of individuals to download and exchange audio and visual files. For many years, the music industry enjoyed growing sales, initially of records and subsequently of CDs. Although there once was an option of purchasing a record with just one or two songs, the music industry quickly abandoned that strategy and replaced it with long-playing vinyl discs and eventually CDs with many pieces. The production costs, overhead, and profits associated with these albums were favorable to the industry, and consumers had no other option than to pay the greater price for all the music on the album. However, as the capacity for downloading files from the Internet expanded, consumers began to download files of single songs and distribute them widely. This was, of course, a violation of copyright, but recording companies were at a loss to prevent it. The recording industry was able to identify some of the major perpetrators of this crime and filed suit against them, demanding payment of substantial fines. Since many of those engaged in such activities were young people who used the Internet portals supplied by the colleges and universities they attended, some administrators of those institutions assisted the recording companies in their investigations and legal proceedings. Other universities attempted to establish a business agreement in which an annual fee was paid to the industry to allow limited downloading, thus providing some protection for the students.

The standoff between the industry and consumers reached a stalemate for several years. However, Apple Computer Company came up with a different business model for the distribution and sale of recordings. In 2001, it launched a free service called iTunes, which allowed customers to download songs. The iTunes software is free, but customers pay a small fee ranging from $0.69 to $1.29 for a specific piece of music. The revenue generated by these fees is paid as a royalty to the copyright holder. Originally, iTunes operated only on Apple equipment, but in 2003 it was modified for use with Windows and other operating systems. Also, iTunes has expanded to permit downloading a wide variety of digital files, including movies, TV shows, podcasts, and other applications for Apple products. The software enables the downloading and management of files for storage and retrieval on computers, laptops, notebooks, cell phones, and handheld devices. The fees for downloads can be paid with a prepaid account or credit card through Store iTunes. The ease of use and low cost quickly turned iTunes into the world's largest online music store.

The major development resulting from the establishment and growth of iTunes was not only the substantial profits accruing to Apple, it was also the creation of expectations among consumers of being able with relative ease to obtain digital files of audio and visual recordings whenever it suited them. No longer did customers have to listen to recordings or view videos at any specific time; they could download these files and enjoy them at any time. Furthermore, most of these files could be stored for delayed and repeated enjoyment. Apple Computer and iTunes have taken the lead in creating a new set of expectations among consumers for on-demand access to entertainment, services, and products. In doing so, they have opened a new and potentially huge expansion of the retail industries. However, Apple is not the only player in this new and rapidly changing business of supplying downloadable digital files on demand.

Again, a brief history will put this development into a broader perspective. In the years before television became widely available, radio developed into a national form of entertainment. Many people had their favorite radio programs, and they scheduled their work and household activities around weekly broadcasts. Families regularly gathered around a radio on weekends or evenings to enjoy such popular shows as *Jack Benny, Edgar Bergen and Charlie McCarthy, This Is Your FBI,* and *Fibber McGee and Molly.* As television broadcasting grew in the 1960s and 1970s, it replaced radio shows as a scheduled event for regular viewing. Family members and coworkers frequently engaged in extended conversations discussing characters and plots. Individuals and families adopted different shows and arranged their schedules to ensure they would be available for viewing. The announcements made by the major television networks each fall detailing the prime-time schedules and shows were eagerly anticipated, and many planned their personal schedules to include blocks of time reserved for their favorite shows.

In later years, videocassette recorder (VCR) systems and digital video disks (DVD) became available to record programs on tape and compact disks. Also, TiVo systems saved programs on computer hard drives. All these devices allowed viewers access to programs anytime after the original telecast. Also, one could subscribe to an Internet service that would allow subsequent viewing of previously broadcast programs. Companies such as Blockbuster, Inc. had many retail distribution stores where one could select tapes or DVDs to rent and view at home for one or more days. All these capabilities for delayed or repeated playing or viewing of programs began to erode the practice of fixed times for listening to radio or viewing television. On-demand entertainment replaced the practice of regular and collective experiencing of entertainment. And in doing so, it spawned a new and huge retail business of downloading digital files of music, movies, television shows, and podcasts.

The company that appears to be dominant in this new enterprise is Netflix, Inc. Established in 1997 in Los Gatos, California, Netflix first offered mailed DVD rentals and then switched to downloaded streaming videos for a flat monthly fee. With a Netflix device, the streaming videos can be run on any home television set. By 2013, Netflix had more than 36 million subscribers, reported over 3 billion in revenue, and had more than 2,000 employees. It offered films and TV programs from all the major and minor production studios to subscribers in North and South America, Scandinavia, and the U.K. Of course, the key to succeeding in this new business is to gain access at reasonable costs to all titles that are copyright protected, and Netflix for a time dominated in this endeavor. On demand downloading now controls this market. Blockbuster, a long standing video and DVD rental company, was forced into bankruptcy. Undoubtedly, the technology used for distribution of entertainment files will change again in the future, and other companies and methods will emerge and become dominant.

Houses and Cars and More

Another niche area of consumerism that has been affected by ICTs is residential real estate. In the past, sales of these properties involved the potential buyer inspecting firsthand the intended object of the sale. Real estate agents took potential buyers on physical tours to see the inside and outside of the available properties within a designated area or neighborhood. Much of the success of the endeavor depended on the agent's ability to perceive the customer's preferences and priorities and to demonstrate the relative value of each available property. The success of the endeavor also depended on the nature of the relationship that developed between the agent and the aspiring buyer. Sometimes the search was brief, and a bid was quickly offered through the agent to the seller. Often the search would be extended for months or even years before an offer was tendered. And frequently, the process would end with no bid or sale. The number and duration of the trips to potential properties were usually determined by the stamina of both the seeker and the agent.

Now the capabilities of ICTs facilitate the listing and viewing of many properties, transforming the processes of real estate visits and sales. Most real estate agents now use the Internet to list and, in many cases, elaborately display with photos and videos the features of residences for which they serve as brokers. Customers perform preliminary searches to screen listed properties before actually visiting them. Furthermore, an increasing number of homebuyers use the Internet to find and examine available properties without the assistance of a real estate agent. Although reliable data are not available, many agents report an increase in residential property sales that are completed without any input from realtors. Many buyers and sellers prefer to bypass the agents, thereby eliminating their fees and reducing the costs of the transaction. It is not clear how widespread the practice of residential sales without a realtor involved will become, but it is certain that the dynamics of that relationship are changing.

Automobile sales are another retail transaction being altered significantly by customer use of the Internet. Prior to the appearance of ICTs, many consumers maintained ongoing relationships with salespersons, and long-term loyalty to both the manufacturer and dealership was common. Negotiations between dealers and customers were frequently extensive and revolved around many details of the automobile model and accessories as well as the method of payment or financing. In most cases, the specifics of dealers' costs and manufacturers' recommended prices for models and accessories were not known to the customers. Many different factors entered into determining the final sale price, and dealers had a great deal of flexibility in realizing their profits. The basic cost of the vehicle and that of different accessories were sometimes gathered in a package price. If the sale was financed over time with the dealer, additional cost variables included the interest rate, the term of the loan, and a down payment. In recent years, many customers have been attracted to leasing vehicles for a fixed monthly rate for multiple years and usually a down payment. Leases add additional factors that can further cloud the actual price paid by the customer and the profit made by the dealer.

Although information about manufacturers' recommended prices has never been kept secret from customers, it was rarely obtained easily. However, with the growth of the Internet, such price information can readily be accessed by anyone with the know-how and the correct website addresses. Furthermore, prices of all models of vehicles as well as all accessories are readily available as are shipping charges from the point of origin to the local dealers. A customer can now approach a dealer armed with all the relevant price information and enter into negotiations concerning the profit margin on the purchase. Furthermore, a customer can visit a nearby dealer and test-drive a vehicle and then shop online among a number of nearby dealers to search for the best price and financing arrangements. The chemistry of the relationship between the dealer or franchise and the buyer now plays a lesser role in the automobile sales industry, and the Internet is the major factor in producing that change. In addition, many dealers now create websites that are very much like entertaining advertisements with audio and visual materials touting the appeal of different models and accessories available for sale. Some of the advertisements include a series of videos featuring celebrities in mini-plots highlighting the cars. Frequently referred to as *webisodes* (web + episodes), these packages are prepared by the manufacturers and then displayed on the websites of the local franchises. Webisodes are, in fact, increasingly used in the promotion of a wide variety of products and services.

The list of retail sales activities that are changing as a result of the use of ICTs is indeed very long, and it is growing continually. Ticket sales for such disparate events as theatrical performances, concerts, and athletic events are all now frequently accomplished online. Personal services such as student tutoring and career counseling are increasingly being done online, and they often involve an offshore setting. The travel industry has undergone substantial change as well. Local travel agents use the Internet to help customers plan trips and make reservations, but travelers are increasingly making their own arrangements online. Major airlines offer discounts to airfares booked online, and many resorts buy email address lists and send enticing advertisements to potential customers. One can look at and buy flowers, holiday decorations, even holiday trees, online and have them shipped long distances, for example, from Alaska to California. Prostitutes are reportedly using the Internet to advertise themselves and negotiate prices, durations, and services to a larger pool of potential customers. In many areas, such solicitations are illegal, but they may be difficult to detect and prosecute.

Perhaps the strongest indicator of the centrality of the Internet in promoting consumerism is that the Nielsen Company is expanding its data collection activities on the patterns and usage of various Internet websites. Nielsen has measured media uses since the early days of radio. During the 1950s, it became the major measuring company of television viewing. Using a variety of data collection methods, Nielsen is accepted as the authority on television viewing, and the results of its surveys and reports drive the revenue streams from advertisers on the major TV networks. Recently, Nielsen announced its plans to expand its data collection and reporting activities to include the Internet, and this is additional evidence that ICTs are exerting greater influences on consumerism.

POST-CONSUMER ECONOMY

As a final brief note to this chapter, it is appropriate to point out that an older economic perspective is currently gaining increasing attention. It speculates that any consumer economy will eventually deteriorate. This is not a new idea, for many people, going all the way back to Aristotle and later Malthus, had written about the perils of uncontrolled growth. However, the resurgence of current interest can be attributed to Garret Hardin, the late professor of human ecology at University of California, Santa Barbara. His 1968 article in *Science* magazine, "The Tragedy of the Commons," argued that both the economy and the environment could not be sustained in an overpopulated world. Several years later, Dennis Meadows, emeritus professor of systems management at the University of New Hampshire, and his colleagues at MIT published *The Limits to Growth*, a report of a computer simulation model that examined the then-existing exponential growth of population and industrial consumption of natural resources. They concluded that unless some constraints were applied to these trends, an economic collapse with devastating consequences would occur.

Recently, several influential economists have proposed that the economies of the developed nations are rapidly approaching the limits of sustainable consumption. They argue that markets need to be created and expanded in other sectors, such as social services and environmental protection activities. Indeed, some claim that the consumer economy has already started to disintegrate. For example, they point out that governments are unable to create economic policies that can remedy the global recession that has rocked international markets since 2008. Economists Larry Summers and Jeff Immelt are quoted in the *Financial Times* newspaper as promoting the idea that the U.S. economy should be more export-oriented and less consumption-oriented; more environmentally-oriented and less energy production-oriented; more bio-, software-, and civil engineering-oriented and less financial engineering-oriented; and, finally, more middle class-oriented and less oriented to income growth that is disproportionate toward a very small share of the population.

These concepts are controversial because some economists believe that existing markets will never be sufficient to reorient the economy and reward an altered labor force. Furthermore, the reorienting of all infrastructures, such as education and training, debt financing, and incentive systems, necessary to support a major expansion of the economy would require massive political support. Given the recent difficulties of enacting any changes in the economic policies of governments in North America and Western Europe, it seems unlikely that a post-consumer economy will receive adequate support among elected officials. Nevertheless, it is interesting to speculate what role, if any, ICTs might play in facilitating such an adaptation to economic and political structure in society. Recall the various transformations of the political and economic systems discussed in Chapter 2. A combination of the informational state as described by Professor Sandra Braman and the post-consumer society might be the next step in societal evolution. Viewed from that perspective, it might be productive to examine various policy implications for strengthening the economy to bring about greater stability.

This chapter has reviewed examples of the new means of buying and selling and new products and services made possible by ICTs. The variety and extent of these innovations in the retail and service industries illustrate clearly that the growth of consumerism in the economies of the developed nations is dependent upon the capabilities afforded by ICTs. Consumption of goods and services supports economic growth and prosperity, and ICT systems enable, facilitate, and encourage ever-expanding markets. Just what role ICTs will play in an informational state and a post-consumer economy remains to be seen.

8 Distance and Time

ALTERED PERCEPTIONS

Information communication technologies (ICTs) are dramatically changing how people now experience both distance and time in their daily lives. Throughout the recorded history of Western Civilization, if not the history of all humankind, most people traveled only short distances from the place of their birth. There were mass migrations, explorations, and religious crusades, but these were the exceptions. For the most part, people spent their lives in narrowly circumscribed areas. Recent analyses of archeological artifacts and DNA samples have established that thousands of years ago, there were large migrations from the Middle East regions north to Europe and Scandinavia. Subsequent migration patterns extended into Eastern Europe and then to Siberia and eventually into what became known as the New World. Hunting and gathering tribes made most of these migrations.

When these tribes, usually extended families, exhausted their local food supplies, they moved into nearby areas. Migrants from the Middle East did eventually travel great distances, but in small segments over many generations. These migrants also brought with them new and developing technologies related to the knowledge, skills, and tools of agriculture and animal husbandry. Nevertheless, from the days of the hunting and gathering tribes up to the eighteenth century, most people had limited geographical mobility. As a consequence, awareness of other peoples, cultures, and environments beyond one's immediate location was at best limited.

Since the Industrial Revolution and into modern times, large masses of people have moved from Europe to North and South America. In the past century transportation systems, particularly railroads and aviation, have expanded enormously. Domestic and international travel is now frequent and extensive for many people in developed nations. All this travel has changed fundamentally our conceptions and experiences of distance and space. And these changes have occurred during the second half of the last century, a relatively short period of time. Of course, the expansion of the mass media has further informed people of other places and other lifestyles. But the recent expansion of ICTs is causing the most dramatic changes.

A corollary of our altered perceptions of distance is a concomitant modification of our experience of time. Space and time are, of course, related in the world of theoretical physics, but within the scope of this investigation, they are linked by the way they are perceived by the users of ICTs. Computers and networks make it possible to be in immediate and continuous contact with people or groups anywhere on the globe, regardless of the local time of the participants in the communication. Those contacts may be synchronous, at the same instant, or asynchronous, at different times. Under such conditions, time takes on a new and very different perspective. Until the recent proliferation of ICTs, the possibility of conducting frequent and extended conversations with people located anywhere on the globe was considered to be the realm of science fiction. Today such communication patterns are routine in developed nations; and with mobile cell phones connected to expanding networks, they are becoming increasingly commonplace in developing countries as well.

To preclude possible confusion in the following discussion, several comments concerning the terminology used are in order. The terms *social distance* and *social time* refer to the changing ways that humans experience and use distance and time in their daily activities and in their perceptions of their world. Over many millennia, these experiences and perceptions have been altered as a result of the changing patterns of observation and activities and the surrounding technologies. The rates of these alterations have escalated, particularly in the past 50 years, as a consequence of the expanded capabilities of both the mass media and ICTs. Therefore, reference is made here to *edistance* and *etime*, denoting that both these dimensions of human experiences are qualitatively different because they take place within the ICT environment. In this discussion, social distance and social time refer to human experience prior to the appearance of ICTs; edistance and etime refer to these dimensions of human experience in the post-ICT era.

The rest of this chapter is divided into two sections. The first explores the history of social distance in human communications from the earliest types of messages, to letters transmitted via postal systems, to the invention of the telegraph and the telephone, and on to contemporary emails, instant messages (IMs), and social networks. The first section also examines the history of social time from the use of the earliest sundials, to the invention of clocks and other timepieces, the establishment of global time zones, and contemporary virtual time simulations. The final section examines the explosion of virtual communities, including: IMs; emails; chat rooms; video sharing sites such as YouTube; social networks such as Facebook, MySpace, and Twitter; and virtual worlds such as SimCity, Second Life, and the World of Warcraft. All these communities incorporate new experiences that integrate virtual perceptions of both edistance and etime.

HISTORY OF SOCIAL DISTANCE AND SOCIAL TIME

For the many thousands of years of pre-literate human existence, the distances involved in social communications were limited to the audible range of the spoken or shouted word, or perhaps to the visual range of fire or smoke signals. However, when humans created writing, they altered communications in two profound ways. First, it became possible to write a message and have the content delivered reliably over much greater distances. Second, written materials could be stored and subsequently recalled for later use. The latter enabled the preservation of knowledge and culture and its transmission to subsequent generations. Both these changes were profound and had implications that are still being played out today in the world of ICTs. They initiated patterns of social communication that altered both distance and time by enabling information to be transmitted over greater physical spaces as well as stored and retrieved for future use.

Development of Message Delivery Systems

After the development of reading and writing capabilities, systems for sending and receiving messages carved on clay tablets or papyrus scrolls were created to circulate records and messages. The technology for circulation followed on the heels of literacy technology, and the geographical perspectives of distance and space expanded for many people. Around 2000 BCE, the Egyptians began developing what today would be recognized as a postal delivery system for sending messages. There is evidence that the Chou dynasty in China had a similar system for delivering official government documents by 1000 BCE. Also, Persian Empire records indicate an operating postal system around 600 BCE. The Roman Empire had the most extensive road and mail delivery systems of the ancient world that included the entire Mediterranean region and most of Northern Europe. The system used horseback riders who reportedly could travel up to 170 miles in a 24-hour period. During the Renaissance, when commerce and trade expanded, especially in Italy and France, private mail delivery services thrived, and many of these companies survived until the late nineteenth century. By that time, most of the public and private mail services in Europe had been taken over by the various central governments.

In the U.S. public mail services were established as early as the seventeenth century in the New England and Middle Atlantic colonies. Immediately after the Revolutionary War, Benjamin Franklin was appointed the first postmaster general in the newly established republic, and he created a series of post roads that enabled a new federal mail delivery system. Many of those early post roads are still in service today, and they formed the foundation of the extensive interstate highway system created in the latter half of the twentieth century.

Following the practice of the European postal systems, Franklin instituted the flat rate concept. Most of the expenses in operating a postal delivery system were administrative and did not reflect increased costs due to the distance from the point of origin to the delivery location. Therefore, the same charge was applied to a letter that was sent to an address anywhere in the country. This concept of the same charge regardless of distance was carried over to the flat rate systems for telegraph and telephones. Reliable and cost-effective communication delivery systems, employing ICTs, are critical to support the ever-expanding economies of both the developed and the developing worlds.

The invention and rapid deployment of the electric telegraph presents another interesting segment of the history of the progression of the transformation of perceptions of time and distance. Developed simultaneously in England and the U.S. during the nineteenth century, the telegraph immediately proved to be superior to a postal delivery system, especially for commercial purposes. Telegraph messages could be sent over long distances, and transmission and receipt occurred with little or no perceived elapsed time. In London the telegraph was used to apprehend a criminal who escaped on a train from the scene of a bank robbery. A telegram containing a description of the perpetrator was quickly dispatched, thus enabling the police to arrest the thief when the train arrived at a distant station. The telegram traveled faster than the train, and the incident helped establish awareness of and appreciation for the efficacy of the telegraph.

An equally symbolic event occurred in the U.S. in 1861 when railroad tracks joining the East and West Coasts were connected near the Continental Divide. It is a lesser-known fact, but also noteworthy, that telegraph lines were placed along the railroad tracks, and the first coast-to-coast electronic connections were completed at the same time. In fact, the spread of telegraph connections throughout North America far outpaced the expansion of the railroads. Suddenly, it was possible to communicate instantly with persons in far distant cities. The use of the telegraph for both personal and commercial purposes further altered people's perceptions of distance. As a symbol of change in communication systems, the New Delhi based newspaper, the *Times of India*, reported on July 15, 2013 that after 162 years of service, the last telegram was sent from the national service. ICTs including email, IMs, and mobile phones made the telegraph service obsolete.

The invention and distribution of the telephone in the late nineteenth and early twentieth centuries probably did more to influence most people's altered perceptions of distance than any other prior message delivery system. The telephone was developed from the technology underlying the telegraph. These two innovations began the era of edistance, the period when sending messages electronically, either with Morse code or by voice transmission, allowed large numbers of people to experience communications over great distances. Although there are competing claims about the telephone's inventor, Alexander Graham Bell received the first U.S. patent for it. It is not widely recognized that Thomas Edison played a major role in developing the switchboard that allowed many telephones to connect to others simultaneously. Today, digital packets of telephone call data are transmitted by copper wires, fiber optic systems, radio signals, and satellite relays. The ease with which people can talk to others who are located far away has changed our perceptions of both the size of the globe and the extent to which the lives of all humans are interconnected. Telephone connections with the Internet have enabled voice and video communication. A system called Voice over Internet Protocol (VOIP) is available at no charge to anyone with Internet access. Two people in opposite positions on the globe can talk and see one another over the Internet. When families can readily view and chat with relatives at great distances, the environment in which children grow up is vastly different from the one that was available only a few decades ago. Distance becomes a different experience.

The most recent and comprehensive innovation enabled by ICTs that affects our experience of distance is the cell or smart phone, which blends the functions of a computer and a telephone to produce a small handheld unit capable of multiple functions. In addition to making and receiving telephone calls, smart phones can incorporate the following functions or applications: send and receive emails, send and receive text messages, maintain appointment calendars, store addresses, access the Internet, play games, watch TV programs, listen to downloaded music, create and send photos and videos, obtain and transmit geographical positions, and as mentioned in the previous chapter, purchase a wide variety of goods and services. All these functions can be achieved from any location that is within the range of the smart phone service provider. Ongoing developments in these phone technologies are expanding the range of coverage and the applications. Mobile phones proliferate at a greater rate than computers, laptops, and notebooks. Since they are much less expensive to produce than computers, their use is increasing, particularly in developing nations.

Changing Perceptions of Time

The ever-changing perceptions of distance over the many thousands of years of recorded history parallel a similar pattern of alterations in how humans have perceived and used time to structure their activities. In prehistoric times, people must have been aware of the near regularity of the amount of daylight and darkness. Daylight allowed people to accomplish the many activities necessary to sustain a family or community, and the periods of darkness limited those activities and were used for rest and recuperation. Fire frequently provided some light, but it was of limited utility for extended periods of time and could be dangerous. Over the millennia, various forms of artificial light were created, including oil lamps, candles, and gaslights. However, it was not until the invention and widespread distribution of the incandescent light bulb in the early twentieth century that people could experience enough artificial light to enable extensive activities during the hours of darkness. The light bulb has made possible the experience of the 24-hour day. In fact, the light bulb may be the most profound technological innovation in history in terms of altered patterns of human behavior.

Early humans in both the Northern and Southern Hemispheres must also have been aware of the changing weather patterns from the heat of summer and the cold of winter. They must have also noticed that the periods of light were longer on the warmer days and shorter on the colder days. The extremes were more drastic among those people who lived greater distances from the equator. As agriculture and animal husbandry gradually replaced hunting and gathering, the recognition of changing weather patterns gave rise to the identification of optimal times for the management of crops and domestic animals. Such observations and behavior patterns promoted the creation of calendars to denote the passing of time in increments greater than one cycle of light and dark. The history of calendars is fascinating, but is far too large a topic to include in this discussion. However, from the perspective of the history of social time, calendars have been instituted and altered mainly by various religious groups to correspond with past events that are significant in their traditions and faiths. The many revisions and alterations of calendars were made to achieve a closer congruence between them and the naturally occurring cycles of the rotation of the Earth on its own axis, the rotation of the moon around the Earth, and the rotation of the Earth around the sun. These cycles correspond closely to our concepts of days, months, and years. The calendar most widely used today in Europe and the Americas is still not completely accurate in terms of those rotations. Witness the awkward adjustments accompanying a leap year.

Humans have been quite ingenious in devising instruments to measure time. From the early days of the Egyptian empires, merkhets or sundials have been ubiquitous. However, they are useless if the sunlight is blocked by a cloud cover. Likewise, instruments for measuring the position and movement of the North Star can be used to measure time, but this technique is viable only when the star is visible, when skies are clear at night. Various other devices, including clepsydras or water clocks, calibrated candles, and different mechanical clocks with springs, were invented and widely used in Europe and elsewhere. In 1581, Galileo noticed that the length of time consumed by a complete swing of a pendulum was fairly consistent, and he devised a clock based on this motion. During the Middle Ages in Europe, a plethora of pendulum and mechanical clocks was constructed in public venues. They were frequently placed in cathedrals and used to activate the bells placed in a nearby tower. It then became possible for people to hear the bells signifying different times of the day and night, usually summoning the clergy and the laity to prayers. Everyone within range became cognizant of the time of day and night and used the bells to arrange their daily activities.

An interesting story related to the awareness of time involves the invention of a maritime clock that could accurately tell time and thereby the location of an ocean-going ship. As maritime traffic for both commercial and military purposes expanded from the fourteenth to the eighteenth centuries, countless shipwrecks caused enormous losses of sailors, vessels, cargoes, and fortunes. The problem was that there was no precise way to determine a ship's location on the high seas. Naval histories are replete with harrowing stories of massive errors in determining a ship's location and the accompanying wreckage disasters. Two pieces of information are necessary to set the specific location of any object on the surface of the globe: latitude and longitude. These two coordinates specify the north/south and east/west location, respectively, of a ship at sea. The specification of latitude, or the north/south position, is a relatively simple procedure that involves measuring the location and angles to the horizon of a number of different celestial objects, for example, the sun or the North Star. However, the measurement of longitude, or the east/west position, is not a simple procedure; it depends upon a series of repeated measurements of the movements of various planets. Inclement weather conditions preclude the necessary visibility. What was needed was an accurate clock or a marine chronometer. Obviously, a pendulum clock would not function on a moving and rolling ship; a new solution was sorely needed.

In 1714, the British Parliament passed the Longitude Act, which offered a substantial cash prize to the person who could solve the dilemma of measuring longitude. A skilled craftsman named John Harrison developed a mechanical clock that could precisely measure time. By comparing the difference between the time at a current position and the time in London, Harrison's clock on board a ship could tell accurate time regardless of the weather conditions or the pitch and roll caused by the swells of the ocean. Harrison's clock did not use a pendulum, nor did it require any lubrication, which could cause a clock to operate differently depending on temperature and air pressure. Eventually, Harrison's clock was reduced in size to that of a pocket watch, and it was widely used in maritime trade throughout the nineteenth century.

The stories of the measurement of longitude and Harrison's clock are relevant to the analysis of time today, because it provided the foundation for the creation of the system of 24 international time zones. Many of the global travel and communication activities that are enabled by ICTs today reference these time zones. During the second half of the nineteenth century, the expansion of both maritime and railroad traffic created difficulties in schedules, coordination, and regulation. In 1884, an International Meridian Conference was convened in Washington, D.C., to approve the establishment of the 24 global time zones. Participants agreed to establish Greenwich near London as the first zone, and the meridian in the middle of the Pacific Ocean as the last zone. These became known worldwide as Greenwich Mean Time (GMT) and the International Date Line (IDL).

The establishment of daylight saving time has further increased the salience of time in the experiences of people in most countries in both the Northern and Southern Hemispheres. Daylight savings time involves moving clocks ahead or back one hour in the spring and fall. The rationale for this system is to provide more daylight in the summer or warmer months, thereby lessening the need for artificial light.

Humans ingeniously created calendars, clocks, and time zones to help structure their activities. More recently, the development of long-distance air travel, electronic communications, and the expansion of radio and television entertainment have all contributed to a highly structured set of human behaviors organized around time. When traveling by air or making a long-distance phone call, attention must be paid to any difference in the time at the point of origin and at the destination. Many radio and television entertainment programs are now available on any device from any location at any time.

Another dimension of time, mobile computing, has altered the way people structure their activities as a consequence of the increasing use of ICTs. Not since the creation of artificial light enabled by the incandescent light bulb has there been such a fundamental change in the work and leisure-time behavior of people, made possible by networks and mobile computing devices. It is now quite convenient and common for people to stay connected at all times with work activities using a handheld device. Not too long ago, people often had to limit their hours of work to what were normally considered business hours, eight or more hours per day. However, as ICTs have made it possible to conduct business affairs anywhere on the globe at any time, it is now common practice to be available 24 hours per day and 7 days per week.

The impact of this kind of behavior on the dynamics of family interactions and relationships is not yet well-documented. However, when adults are online 24/7, it should not be a surprise that children follow suit and use mobile computing to increase interactions with their peers. Recall from Chapter 5 on eLearning that socialization in multi-generational agricultural families was accomplished by children observing and emulating the behavior of their adult relatives. Adults and children interacting with other adults and children is certainly not a new social activity, but what is new and somewhat troubling is the extent of the interaction among children that is not observable by adults. Communications among children that take place on ICT-enabled systems are for the most part not visible to parents, teachers, or others who supposedly supervise children's daily lives.

Children have always been involved in experimenting with behaviors that adults would censure if they became aware of them. Kids bully others in the corner of the schoolyard, smoke in the woods, or sexually explore themselves and others behind the barn or in another private space. However, what is different in online adolescent behaviors now is that instead of the materials or behaviors being witnessed by a handful of peers, the messages, pictures, or videos that are posted on sites such as Facebook, MySpace, or You Tube are now viewed by an untold number of persons nearby and around the globe. Furthermore, these online postings become a permanent fixture on the Internet, thereby extending their presence and influence. In earlier days, the consequences of adolescent errors in judgment and behavior lasted only as long as the juvenile records existed or people remembered them. The vastly enlarged exposure enabled by ICTs is leading to unfortunate and problematic outcomes, including stalking, emotional turmoil, and suicides. This is a clear example of how ICTs have created a series of problems that are not new in character, but whose implications are vastly complicated by the fact that Internet postings can result in an extension of the half-life of an indiscretion. In such cases, etime takes on a new dimension approaching permanency, and edistance is greatly expanded beyond the small peer group to include any persons with Internet access. Clearly, edistance and etime are qualitatively different aspects of human experience.

VIRTUAL COMMUNITIES

The phrase *virtual community* refers to groups of people who use online capabilities to communicate or to pursue common activities or goals. These groups function for many different purposes. However, they all share the common properties of operating in both espace and etime. With few exceptions, the boundaries of time and space in the physical world are fixed; those same boundaries in virtual worlds are fluid. Involvement in virtual communities occurs without regard to the dimensions of space and time as they are experienced in participants' physical existence. This section examines 10 types of communication capabilities and patterns supporting virtual communities. They differ in whether they employ ICTs synchronously or asynchronously, and with one or multiple participants.

1. Instant messages (IMs)
2. Emails
3. Chat rooms
4. Listservs
5. Blogs
6. Online education
7. Video sharing
8. Social networks
9. Virtual worlds
10. Discussion and image boards

The order in which the forms are listed implies some correspondence to the number of people involved in the communities. These numbers range from a dyad in IMs to the millions who may be involved in video sharing, social networks, virtual worlds, and discussion and image boards. The order of the list also roughly matches the historical order of their creation, distribution, and use. For certain, many more types of virtual communities exist than those listed. Virtual communities are among the fastest growing segments of all ICT implementations. So many new and innovative instances of virtual communities arise practically every day that it is impossible to create and publish an up-to-date list that will not be obsolete immediately. The analyses presented in this section are simply models of the burgeoning field of virtual communities. These 10 categories are not mutually exclusive, for several share common properties. For example, some types of video sharing, such as YouTube, are also social networks. Scientists always prefer mutually exclusive categories, but human behaviors do not always correspond to such schematic patterns.

To provide a framework for examining these 10 forms of virtual communities, Figure 8.1 portrays them in a conceptual space defined by two dimensions, synchronicity and number of participants. The former is defined by the two mutually exclusive categories of synchronous (S) and asynchronous (A) communications. The latter consists of three patterns: single person to single person (SS), single person to many persons (SM), and many persons to many persons (MM). To facilitate the discussion, the six cells in the table are labeled A through F. The positions of the 10 forms within the six cells indicate the categories are not mutually exclusive, and some of these virtual communities can appropriately be placed into multiple cells.

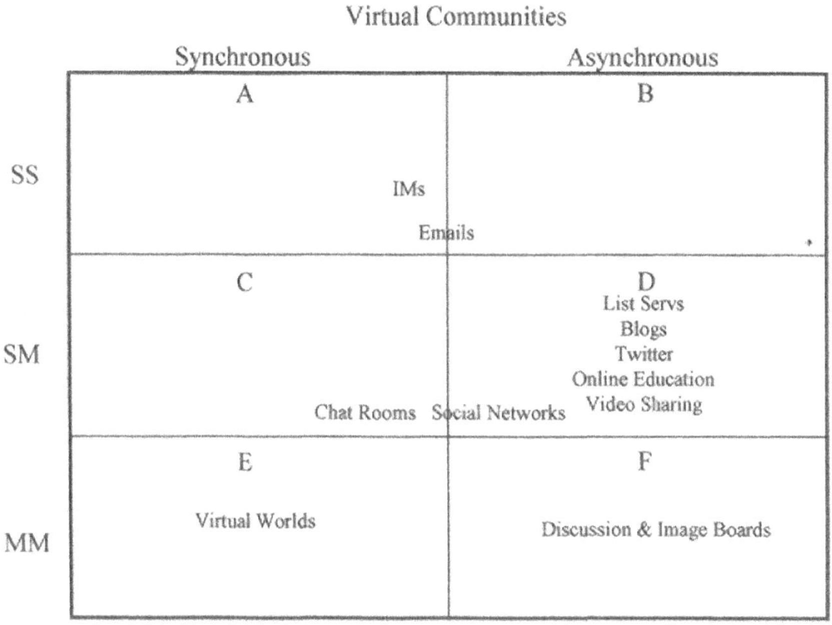

Virtual Communities

	Synchronous	Asynchronous
	A	B
SS	IMs Emails	
	C	D List Servs Blogs
SM		Twitter Online Education
	Chat Rooms Social Networks	Video Sharing
	E	F
MM	Virtual Worlds	Discussion & Image Boards

Figure 8.1 Virtual communities.

Instant Messages

IMs are placed in cell A, because they are usually synchronous (S) exchanges from one person to one other person (SS). IMs first appeared on the early time-sharing systems in the 1960s when only person-to-person messaging was possible. Eventually, systems developed capacities for sending IMs from one person to multiple recipients or "buddies" on distribution lists. Note that IMs are placed in the lower right-hand corner of cell A, reflecting that they are not always responded to immediately. When that happens, the IMs are asynchronous and would be placed in cell B. If the IMs are sent to previously defined buddies, then they should be in cell D, corresponding to the single to many persons (SM) pattern. IMs are placed in the A cell but close to both the asynchronous and SM borders, corresponding to cells B, C, and D.

IMs are used widely now on mobile phones\, and they are enormously popular among adolescents and young adults. One teenager commenting on the lack of technology competence of his parents complained that they were still using telephones to talk to one another. However, older people are fast becoming IMers. Many mobile devices have small keypads that represent both letters and numbers, and they can be awkward to use for sending long messages. Consequently, as mentioned in Chapter 2, a new form of communication has emerged that uses a combination of letters, numbers, and special symbols to send messages. The absence of capital letters and little, if any, punctuation also characterizes this new form of messaging. The members of virtual communities have created these altered forms of electronically transmitted English to decrease the number of pad touches or key strokes necessary to send messages, and they now show up in many other forms of offline communication as well. University and college professors report that papers and dissertations submitted by students today increasingly use many of these abbreviated forms of writing. These emerging forms of abbreviated writing are found in all the virtual communities, but their prevalence is most pronounced in the IMs.

The episode reported in Chapter 2 demonstrated that the initial objections of many faculty members to these new forms of written text are due to the departure from what they perceive as the time-honored principles of effective writing, including correct spelling, punctuation, and grammar. However, once they recognize that it is possible to understand these forms of communication without ambiguity, the resistance begins to diminish. After all, the purpose of punctuation, correct spelling, and proper sentence structure is to communicate clearly. If the message is understood, the traditional rules then become less salient. Increasingly, schoolteachers and higher education faculty are accepting work from students using the new forms of writing as they are emerging from the ICT-enabled virtual communities. They reason that soon, these students are likely to be the editors of all sorts of publications; as long as the content is clearly stated, readers will become more accepting of the new forms of communication. It is also possible that some faculty are eager to appear knowledgeable to their students with respect to the latest practices and fads in the world of ICTs.

Emails

Email appears at the intersects of cells A, B, and D. Email also had an early start, appearing in the 1970s in the U.S. Department of Defense supported project, Advanced Research Project Agency Network (ARPANET). Initially, this was a synchronous system similar to IMs in that both sender and receiver had to be on the system at the same time. It quickly became asynchronous and supported both single (SS) and multiple (SM) activities. The message and header formats of the ARPANET emails have persisted, and today they are still quite similar. By 1986, the National Science Foundation network, known as NSFNET, had taken over the civilian research section of the ARPANET, which eventually expanded into the current Internet, which is now open to commercial activities.

The positioning of emails on the border between the synchronous cell A and the asynchronous cell B in Figure 8.1 reflects that the response is sometimes immediate and sometimes delayed. Furthermore, emails are frequently sent synchronously to many people, a pattern close to the SM format.

Chat Rooms

Chat rooms are the only entry in cell C, and they are placed close to the border of cells C and E. The reason is that there are two types of chat rooms, directed and undirected. A *directed* chat room is similar to a discussion with a group leader, and an *undirected* chat room is analogous to a conversation pit in which participants feel free to talk at any time about any topic that strikes their fancy. A common example of the directed chat room is a group psychotherapy session. This is an online adaptation of the traditional face-to-face group encounter in which a therapist guides the discussion to address issues about which the members are seeking professional assistance, for example, addictions, compulsions, or life course transitions. The group members commit to the sessions and fees in advance; at the appointed time, they log in to the session for a predetermined number of minutes. The therapist has the challenge of directing the discussion without the benefit of observing the behaviors and body language of the participants. Also, the group members are responding only to the audio stimuli of the discussions. Therapists claim that this type of group session can be helpful in dealing with certain kinds of psychological problems. Directed chat rooms are synchronous and fit the SM pattern.

In the undirected chat rooms, the conversations are quite free form and not guided by anyone. In the physical world they are analogous to dinner party conversations. People visit these chat rooms for a variety of reasons, and sometimes they are simply interested in socializing with strangers or possibly making new friends. Many chat rooms are organized around specific topics such as stamp collecting, organic gardening, or folk dancing. Interested people log into the chat room at any time and decide if they wish to participate on whatever topics are being discussed at the moment. Undirected chat rooms that discuss topics related to human sexuality have generated a great deal of controversy. The discussions frequently include content that many consider inappropriate for children. Despite warnings that only adults should visit these rooms and the installation of blocks on computers used by children, the problem persists. Suggestive videos of people available for discussions have appeared online. Some chat room services have established rules to which all who enter the discussion must agree to abide. Violation of the rules will result in expulsion from the site. Monitors are paid to observe the entries, and they have expulsion authority and can remove any content they deem inappropriate and prevent offenders from entering the room. Yet the number of sexually explicit chat rooms continues to grow.

A recent expansion of the chat room category is Snapchat. Designed by a group of Stanford University students, Snapchat adds photos or snaps to the online exchanges. Anyone can post images and specify how long they can be viewed, from one to 10 seconds, and then they are deleted. The most common images posted are "selfies" or self-portraits. Snapchat is used heavily to transmit nude or sexually graphic images, and the users are primarily young persons. Snapshot instantly became a popular application, and it reported that 20 million snaps were viewed per day by the end of 2012. As with several other ICT innovations Snapchat is involved in a legal battle initiated by a person who claims to have originated the application and seeks reimbursement for the theft of intellectual property.

Listservs, Blogs, Online Education, and Video Sharing

The expansion of emails to multiple receivers leads directly to the four virtual communities that are listed in cell D of Figure 8.1, listservs, blogs, online education, and image sharing. The major difference between cells B and D is that in the latter there are always multiple receivers, and frequently there are hundreds or even thousands of receivers. Listservs are essentially emails with multiple receivers. Listservs consist of the email addresses of persons who subscribe because they share common interests or concerns. The number of different listservs is enormous, and they are as diverse as can be imagined. The topics range across the entire realm of human interests and activities, from aardvarks to zygology. Some listservs have been in existence for years, and others may generate little interest and last only a few days. Many listservs are organized and operated independently. A group of people who share a common interest subscribe, and volunteers handle the tasks of sending emails and maintaining the membership lists. When the size of the list becomes too large and the administration becomes too burdensome for the volunteers, commercial services are frequently used. Some of the more popular listservs include Pinterest (photograph sharing), LinkedIn (professional networking), Kickstarter (fund raising), and Reddit (news sharing). The last example was involved in the crowd-sourcing debacle following the Boston Marathon bombing.

Listservs are often used to distribute various kinds of materials from one source that will be of interest to list members. For example, farmers who raise soybeans in the U.S. Midwest may want to share information concerning strategies for obtaining optimal crop yields or information on projected prices, pest control, or new strains of seeds. List members who come across materials that will be of interest to others will email a copy to the list manager, who in turn will send it out to all other members. Not too many generations ago, farmers would share this kind of information in a conversation with an agricultural agent during an annual county fair or an informal face-to-face meeting in a local general store. Such information exchange groups exist in multiples of thousands and are sometimes known as forums, message boards, or discussion groups. ICT-enabled listservs greatly enhance the frequency and salience of such information exchanges.

Blogs, the second entry in cell D, are closely related to listservs. The word *blog* is a concatenation of selected letters from the words *web* and *log*. Like listservs, blogs are usually directed by one person, but there are frequently more discussions in the form of responses. Blogs are a website where the blogger posts a message that is then available to anyone who subscribes to or visits that site. Many bloggers send out a message to a list of subscribers that a new post has appeared. Hence, blogs and listservs are both asynchronous and fit the SM pattern. However, listservs tend to focus on the dissemination of information; blogs are more likely to generate responses or comments regarding the values and viewpoints of the blogger and the members of the community. Listservs are comparable to the news sections of the daily paper, and blogs are similar to the opinion and letters-to-the-editors pages.

Blogs became very popular in the early 1990s. Although no one knows for certain just how many blogs exist, an undocumented, and probably exaggerated, estimate put the number at 125 million. Despite the large number of blogs, most seem to have a short life. However, a small number of popular blogs do attract large numbers of readers. In fact, some of the most popular blogs have so many subscribers that advertising agencies pay for space on the website to promote the products and services of their clients, thus generating an income for the blogger. It is quite possible for the small number of successful bloggers to earn a substantial income from such advertising revenues. This recently created form of independent commentary on current affairs has proven so attractive to some senior members of the journalism profession that they have left their employment in the print or television media and struck out on their own as independents with a personal blog.

Many aspiring young writers are attempting to create their own financially rewarding blogs, thus spawning what some are calling *citizen journalism* or *crowd-sourced* news reporting and analyses. Prior to the advent of virtual communities, news reporting and analyses were accomplished by a relatively small number of professional journalists employed by the media companies that produced newspapers and magazines and operated television stations. Virtual communities, and particularly blogs, have now made it possible for more people to engage in news reportage and interpretation. These new sources are free from the editorial constraints that in some cases employers imposed upon journalists. Some claim that the writing in the blogs is inferior due to the absence of editorial oversight. Others claim that the investigative reporting on blogs is superior, because citizen journalists are not influenced by the values or political preferences of the owners and publishers of the traditional media companies. There appears to be some merit to both positions.

Twitter, a variation of blogs, is a short message system that limits postings to 140 characters. Twitter was started by businessman Jack Dorsey in 2006 and quickly grew to 200 million worldwide users by 2013, who post an estimated 350 million tweets every day. Many people use Twitter simply to tell others about their daily activities and feelings. Twitter has gained substantial notoriety as many political leaders and celebrities of the entertainment world use Twitter. A recent event in China illustrates the potential power of such microblogs to inform the public and foment reactions. In 2011 a high-speed train crashed near Wenzhou in coastal China, resulting in 40 deaths and 191 injuries. Several tweets, or weibos as they are called in China, from the scene of the accident alerted many citizens to the tragedy. The messages decried the fact that medical and rescue assistance was so slow in arriving at the scene. Several government agencies reacted by either ignoring or downplaying the news. However, within just a few days more than 26 million tweets were posted criticizing the railroad and government censorship. Officials could no longer ignore the outcry, and several apologies and corrections were issued. Such an event is highly unusual in China, and weibos are the direct cause of the policy changes.

The sixth type of virtual community is the large category of online education. As mentioned in Chapter 5, online education is one of the fastest growing sectors of ICT implementation. It was also pointed out in that chapter that online higher education course implementation in the form of Massive Open Online Courses (MOOCs) is expanding at a rapid rate. Both the traditional non-profit and the newly expanding for-profit educational institutions are increasing their online course offerings, and this growth is occurring at all levels, from elementary to adult and continuing professional development. However, adult education courses that are intended to provide instruction leading to employment or occupational advancement account for most of the growth. These programs are offered mainly by for-profit institutions and community colleges and are typically conducted completely online. One or sometimes several teachers communicate with a virtual classroom of students. Most online courses are taken asynchronously, for the capability of "going to class" at any time of day or night from a computer that is located anywhere is a major attraction for students. An increasing number of educational institutions offer courses and degree programs in which students never set foot on the campus. As mentioned earlier, one can earn a master's degree in engineering from Stanford University by completing only online courses. Online courses are expanding at the secondary school level as well.

Video sharing is undoubtedly the fastest growing of all ICT platforms. The dominant player in this category is YouTube. *Forbes* magazine recently estimated that YouTube is the third most frequently visited Internet site after Google and Facebook. YouTube was launched in February 2005 by three innovative young men who were formerly colleagues at PayPal, a secure online system for processing electronic fund transfers. YouTube was an immediate success, and its phenomenal growth is a hallmark of the potential of the Internet for innovation and rapid dissemination. In November 2006, Google purchased YouTube for $1.65 billion, a value that was attained in just 15 months. The YouTube software enables anyone to upload a video file and make it immediately available worldwide. Two billion videos are estimated to be accessed each day, and that means that advertising revenues are collected on 2 billion YouTube sites. More videos are accessed on YouTube in one 24-hour period than are viewed on the television broadcasts of the three original major networks in North America combined. As one might expect, many of the videos are copies or altered versions of content that is under copyright protection. The controversies generated by YouTube are extensive, complex, and likely to be unresolved for some time.

Social Networks

Social networks are Internet-based systems that allow people to create and maintain contact with others. Their position in Figure 8.1 at the intersects of cells C, D, E, and F implies that the participants in social networks operate at times synchronously and at other times asynchronously. The communications may be initiated and received by either one person or many persons. More than 200 active social network sites exist, and the number continues to increase each year. Some social networks focus on a particular common interest, such as English literature, or a social role, such as young mothers. Others have no specific focus; rather, they grow by locating friends or colleagues and inviting them to join the network. The most popular social networks are Facebook, MySpace, and Friendster. As these have all been introduced in earlier chapters, mention is made here simply to locate them in the two-dimensional space defined by Figure 8.1. It should be pointed out that Facebook has emerged as the dominant social network. Facebook, a privately owned corporation, reports that it has more than 600 million users. Consider that only 10 years ago it would have been preposterous to speculate that one could contact that many persons on any communication system. Facebook's public identity has been energized by a popular and critically acclaimed feature motion picture, *The Social Network*, released in 2010.

Virtual Worlds

Virtual worlds have a variety of meanings in various contexts. In this discussion, the term refers to the many online synchronous programs that engage people in simulating some aspect of a real or fantasy world. Many virtual worlds are games in which players assume one or more simulated roles and accumulate points or prizes or attain certificates of competence or accomplishment. Some of these accumulations can be converted to offline funds. As mentioned in Chapter 5, these games are frequently referred to as massively multi-player online role-playing games (MMORPGs). Hundreds of MMORPGs exist. A 2009 estimate indicated that they annually generate more than $24 billion in revenue. That number has surely grown since then.

In Figure 8.1, virtual worlds are placed in cell E, indicating that in MMORPGs subscribers participate in groups, sometimes with friends, otherwise with strangers met online. Five examples are described briefly to illustrate this category of virtual worlds: Dungeons and Dragons, one of the earliest online games; World of Warcraft, the most popular of the virtual world games; EVE Online, a science fiction fantasy game; Second Life, the most diverse virtual world; and Grand Theft Auto, the most controversial commercial product in this class. Although many more programs simulate virtual worlds, these five demonstrate the variety of ways that distance and time take on new dimensions in virtual communities.

Dungeons and Dragons (DnD), one of the oldest virtual worlds, was first developed as a board game before computer time-sharing systems became available. In the 1970s, an online version for single players appeared on teletype machines. In the latest version, subscribers pay a monthly fee between $7 and $10, depending on the length of their commitment. They then play the role of a character that virtually progresses through various routes or dungeons and encounters a variety of demons and threats to attain ever more demanding levels of skill and competence. Players can operate as individual avatars or as members of a virtual team. When logged onto DnD, individuals experience virtual distance and time, and they are independent of the physical world. This experience of total immersion in DnD reportedly can cause addiction to the game. The current owner of DnD, Wizards of the Coast, has achieved substantial financial success with related products such as books, manuals, figures of game characters, and themed items of clothing. All these objects contribute to the encompassing experience of participating in the virtual world of DnD.

World of Warcraft (WoW) is by far the largest MMORPG in terms of number of subscribers and revenue. It was launched in 1994 and significantly upgraded in 2004 to include a three-dimensional virtual world experience. Its publisher, Blizzard Entertainment, reported more than 10 million subscribers in 2012 in North America, the U.K., Europe, China, and South Korea. Players create an avatar and choose a virtual setting in which to operate. They also choose from a limited number of occupations, join a guild, and through experiences advance their skill levels. They then travel through that setting, usually in teams, engaging in combat with other avatars or overcoming obstacles in the environment. They can accumulate weapons and funds to use in subsequent combat or environmental challenges, such as randomly occurring weather catastrophes. The realistic three-dimensional portrayal of WoW attracts and holds the attention of players, who pay nominal fees for participation. Avatars can be killed by other players or by game-produced fatalities. With sufficient game funds, they can be resurrected and continue playing. A player leaving the game can return at a later time, but a penalty in the form of a fine or diminished resources or skills is usually imposed. WoW encourages sustained participation and supports the experience of playing in an environment that is structured by virtual distance and time. The producers of WoW also sell a variety of products offline to reinforce further the experience of total immersion in being a warrior.

EVE Online is a science fiction MMORPG set in a mythical wormhole named EVE. The action takes place 21 centuries in the future. It simulates conflicts among warring populations of races descended from the humans who long ago abandoned the planet Earth when they exhausted its natural resources. Players join one of the four races and navigate spaceships through a universe of thousands of star systems, engaging in a number of activities including combat, manufacturing, trade, and piracy. The combat may be against the starships of other players or against creatures or obstacles in outer space. Players advance in the game by attaining new skills through training programs. Training is accomplished during both playing and offline time, thus encouraging engagement and longevity. EVE Online is particularly popular in the European nations. CCP Games, the developer and publisher, reported that in 2008, the European nations combined accounted for more than 40 percent of all subscribers. Their average age was 27, and males accounted for 95 percent of the total. The company also reported that players averaged 2.5 hours of daily engagement with EVE Online. That amounts to an average of 17.5 hours per week, close to a half-time job. By 2013 EVE Online had more than 500,000 subscribers.

Introduced in Chapter 5, Second Life (SL) is certainly the largest and probably the most innovative virtual world, for it enables a wide variety of activities that mirror human behaviors in the physical world, but it allows these activities to take place in virtual distance and time. In SL, avatars of players, called Residents, transcend space and fly to and from locations in the virtual world, visiting sites, conducting business affairs, and socializing with the avatars of others who are online at the same time. Linden Labs, based in San Francisco, California, launched SL in 2003, and by 2013 it reports an average of 1 million visits each month. Although precise data have not been released, it appears that the numbers of SL users has recently declined. Linden Labs defines SL as a three-dimensional digital world imagined, created, and owned by those who live there. Although one can role-play games in SL, it is more than a MMORPG. It is a comprehensive digital world enabling a wide variety of activities, including commerce, land development, education, and service provision.

The last example of a virtual world is Grand Theft Auto (GTA). It is among the most financially successful and controversial MMORPGs, for it contains both violent and sexually explicit segments. GTA was first introduced in 1997 and has gone through many expansions, including one that was condemned by a parents group because it contained full frontal nudity. Players travel through a mythical city committing a variety of crimes, including automobile theft, drug dealing, and murder. They accumulate points for success in these unsavory endeavors and advance their status to undertake more challenging assignments. Since its initial launch, GTA producers claim more than 200 million sales. It is a sad commentary on contemporary culture that this virtual community attains such widespread success.

Discussion and Image Boards

Discussion and image boards appear in cell F of Figure 8.1, indicating that they are asynchronous communications that are initiated and received among many different participants. Discussion boards initially contained only text. As with listservs, people who subscribe share a common interest and wish to access updated information relevant to that issue. However, discussion boards encourage more exchanges of both information and personal viewpoints among their subscribers. In that respect, discussion boards share some of the characteristics of blogs. As the capabilities of the Internet for rapid uploading and downloading increased, many discussion boards expanded their content to include both text and visuals, thus giving rise to image boards. Adding photo and video images produced an expanded range of content for these boards, and the number of subscribers who visited these websites increased quickly. One of the most popular of these image boards is 4chan. Initially, it was the production of a Japanese firm that portrayed anime, a popular form of a soft pornography comic strip. It has now expanded to become a vast collection of images promoting a wide variety of different interests, including Japanese culture, arts and crafts, and cuisine. However, a great deal of the content has most recently become some form of sexual solicitation or hard-core pornography.

The social science research community has studied virtual communities for some time now, and the literature is vast and growing at a rapid rate. Community has always been a major topic of inquiry from a sociological perspective, and the interest in virtual communities is a natural extension of that theoretical and empirical work. An organization that facilitates communication among scholars active in this field is the Association of Internet Researchers (AoIR). The organization conducts annual meetings where members present papers and maintains an active listserv with many daily postings from scholars around the world. One of the leaders of this group of researchers is Barry Wellman, professor of sociology at the University of Toronto.

All the virtual communities described here illustrate the ability of the Internet to allow a fundamentally different experience of time and space. From the earliest experiences of humans with limited perspectives of their environments and time cycles to the virtual worlds enabled by ICTs, an enormous expansion of time and space has taken place. It is interesting to speculate how and where the next expansion will further enlarge our physical and virtual worlds. Neuroscientists, who study the relationships between brain wave activity and the emotions and behaviors of humans and laboratory animals, have expressed an interest in expanding their inquiries to include the experience of being in two places simultaneously, the physical and the virtual. This promises to be an intriguing new field of research.

9 Future

The preceding chapters have raised a large number of issues that are relevant to the diverse ways that information communication technologies (ICTs) are influencing patterns of social change, many of which have their roots in ancient history. The theoretical backgrounds for these analyses focused on the location of human societies, the expanding modes of literacy, and the economic and political foundations of social structure, leading to the concept of the informational state as proposed by Professor Sandra Braman. The inquiries in this book have ranged from normative order, to socialization and education, the creation and use of knowledge, consumerism, and experiences of distance and time. In this final chapter, the discussion turns to the future. However, it should be pointed out that as stated in Chapter 1, the intention is not prediction. No attempt will be made here to anticipate how ICTs themselves will develop in the future. Rather, five issues raised in earlier chapters are especially promising to revisit with a view toward how they may evolve in the future and influence further social change. These issues are data visualization, new forms of literacy, the balance between privacy and transparency, culture wars, and the emergence of the informational state in the context of advancing globalization.

DATA VISUALIZATION

As mentioned in Chapter 6 on knowledge, ICTs are enabling the creation and analyses of large-scale databases for scientific and policy-relevant research. The current capacity of ICTs is allowing investigations that previously were considered impractical due to their extraordinary data manipulation and calculation requirements. Furthermore, the enhanced capabilities of ICTs for graphical output and animation support the development of new systems for displaying the results of these analyses of large-scale databases. These new modes of display are now referred to as *data visualization*, and these exciting developments show enormous promise for enhancing public understanding of the policy implications of scientific research.

Statisticians and graphic designers are using computers to develop new methods of summarizing and displaying data from large databases. An excellent example of animated display of data from a large-scale database is the work of Professor Hans Rosling, an international public health expert at the Karolinska Hospital in Sweden. He creates animated statistical displays to portray changing epidemiological, economic, and educational conditions in different nations over time. Using colored displays of various forms to signify data, the graphical animations summarize large series of trend statistics and produce a comprehensive, animated, international picture that is intuitively both informative and appealing. This work is indicative of possible future applications in a wide variety of substantive settings.

Under the direction of Professor Benjamin Shneiderman at the University of Maryland, scientists have developed an analytic display known as *tree mapping*. Using various colored and sized rectangles, the displays portray a wide variety of large-scale data sets. Tree maps have been used to chart performance data of many organizations and individuals in multiple dimensions. The intention of such visualizations is to enhance analyses of large complex data sets, and they provide new ways for scientists to portray the results in animated forms.

It is interesting to speculate what future applications merging large-scale databases and dynamic displays might produce. Some projects already underway may give us some basis for speculation. Scholars at the University of Virginia have for some years now been creating a three-dimensional computer model of Rome in the year 325 CE with the intent that it will be useful for both research and instructional purposes. A similar model has been created of the Appalachian Valley during the time of the U.S. Civil War. Another project uses historical data and photographs to recreate the views of the Battle of Gettysburg during the same war. A different project relies on dynamic displays to observe the incidence of trials of alleged witchcraft in colonial Salem, Massachusetts. Such maps are increasingly referred to as Spatial Humanities research projects. Would such replicas be valuable for other historical settings? Might it be useful to incorporate some of the features of virtual worlds with avatars to interact in ancient settings? How would one assess the heuristic value of such dynamic displays based on large-scale databases?

NEW FORMS OF LITERACY

As mentioned in previous chapters, several new kinds of literacy appear to be emerging. First, instant messaging (IM) has spawned a style that omits all punctuation and capitalization, including for personal names. This style also extensively uses abbreviations, phonetic spellings, and concatenations with numbers. Also, recall the reaction of the professor to the student who used this form of literacy in her doctoral dissertation. The professor realized that he was most likely witnessing the evolution of a new form of scholarly communication.

A second, and perhaps more fundamental, form of new literacy is *iconography*. When IBM introduced its personal computers in 1981, users typed commands to execute instructions and programs in an operating system developed by Microsoft, then a new corporation. Most early personal computer users spent an inordinate amount of time developing the skills required to write error-free instructions. When Apple introduced the Macintosh personal computer in 1984, its operating system was a graphical user interface (GUI) that employed a mouse to direct a cursor on the screen to pull down menus. The options for the menus were small images representing the available functional programs. The user did not have to learn and memorize the correct format and spelling of the instructions. Rather, moving the cursor over the image and selecting it by clicking became the mode of operation. The success of the graphical user interface was immediate, and all personal computer manufacturers today use point-and-click operating systems. The images are referred to as *icons*, a word taken from the religious paintings that were common in early Eastern Orthodox Christian cathedrals.

Icons are increasingly being used for the control systems in laptop and mobile devices. Microsoft introduced what it called a "coffee table computer" that used only icons to activate various functions. There was neither a mouse nor a keyboard. The user simply touched the icon with a finger or a stylus. Many of the mobile phones and newer Apple products use touch screens for most controls. Users touch or swipe their fingers over icons to select, activate, or change application programs. One touch on a screen may activate literally hundreds of instructions to implement many programs. The applications that can be purchased for an Apple iPhone, for example, now number in the thousands. The language of the icons is the primary, if not exclusive, mode of communication with the hardware and software of an increasing number of mobile ICTs. The latest version of the Microsoft operating system, Windows 8, was released in 2012. It was designed to operate with an interface to compete with iOS, Apple's operating system for mobile devices, which has neither a keyboard nor a mouse. Touching or sweeping icons on the screen activates applications or programs. Windows 8 operates on desktop computers, laptops, and notebooks as well as handheld devices. Iconography is becoming common as the primary human-to-computer interface (HCI).

Common definitions and icons of various functions such as print, save, call, or buy are emerging across different ICT platforms. In retrospect, it is clear that the innovation of the GUI with the first Apple Macintosh computers might have ushered in an important expansion of literacy. Many are calling this the next stage in the development of literacy that dates back to the clay tablets of the Sumerians and the hieroglyphics on the papyrus scrolls of ancient Egypt. As with earlier innovations in the development of literacy, icons will not replace text. Rather, iconography will likely complement existing formats and further enrich human capacities for communication.

PRIVACY AND TRANSPARENCY

Another issue that emerges from the analyses of ICTs and social change that will be a major focus in the future is that of the balance between privacy and transparency. In this new era of widespread use of ICTs, an appropriate balance is needed between an individual's or an organization's need for privacy and autonomy on the one hand and society's need for transparency and accountability on the other. As mentioned in Chapter 4 on normative order, privacy is a critical value in Western Civilization, and it has taken on new and important dimensions as ICTs have proliferated in society. In the first 75 years of the existence of the U.S. as an independent republic, privacy was a readily acknowledged right. In the Bill of Rights appended to the U.S. Constitution, the right of privacy simply guaranteed that no government or private agency could invade the domicile of any citizen without due process or prior court approval. Unless suspected of some criminal activity, each person had the right to protect his or her personal privacy against the intrusion of any other person or government agency. That right to protect one's own privacy has withstood many court challenges over the years, and until recently it has been clearly applied. However, that principle has been reinterpreted since ICTs enabled the collection of information about many people's private lives. The problem becomes acute when those accumulated data are made available for a substantial fee to both public and private agencies. Frequently, the disclosure has been accomplished without the knowledge or prior approval of the individual.

As discussed in Chapter 7 on consumerism, retailers have strived to increase sales and profits that have driven the accumulation of these data and the attendant practice of compromising privacy. The accumulation of individual information on shopping patterns has blurred the distinction between privacy and ownership. It is not clear who owns the accumulated data files that identify individuals and their online behaviors. Nevertheless, the companies that track such data claim ownership, and they sell the information to advertising agencies and retailers. Public opinion polls show that some people strongly object to such practices and consider them direct violations of their rights of privacy. Several class action lawsuits are underway to stop such practices, and it is by no means clear what the outcome of such cases will be. The same opinion polls show that large numbers of people are not in the least bothered by the accumulation and distribution of such personal information. However, it seems likely that in the not-too-distant future, Social Security numbers will be used to link data on patterns of Web surfing and online shopping to other files containing personal information from medical records, criminal proceedings, motor vehicle code violations, and real estate or other financial transactions. When this happens, further objections to such invasions of personal privacy are likely.

Another dimension of this issue concerns the rights of organizations, both private and public, to withhold information about their operations and finances. Some people claim that government agencies are obligated to share with taxpayers all information about operations and finances. After all, such agencies operate for the common good of the citizens, and the officials serve at the pleasure of the public. Likewise, publicly owned corporations are obligated to keep their shareholders similarly informed. Other people agree in principle with such transparency, but they claim that in practice both public and private organizations need to keep certain information private to ensure security for governments and profitability for corporations. This delicate balance between privacy and transparency has operated reasonably well in the past. Scandals in both public and private organizations occurred when secrecy enabled some individuals to behave improperly to their personal advantage. Nevertheless, the extreme swinging of the pendulum seemed to provide the necessary corrections, and a working balance was maintained. However, ICTs and their capacity to accumulate and store all kinds of information are changing the dynamics of that balance. Experience has proven that no data file is completely secure regardless of the protective barriers. Hackers have convincingly demonstrated that any data storage system can eventually be breached. Furthermore, when data files are compromised and released to the general public, ICTs and the mass media can spread that information worldwide in a matter of minutes.

A notable recent example of disclosure of sensitive data includes those activities carried out under the auspices of the WikiLeaks Foundation. Thus far, it has compromised data describing recent military operations in the Middle East and diplomatic exchanges among many global political leaders and organizations. Furthermore, WikiLeaks has announced that it plans to release data revealing the international financial operations of major banks in Western Europe and North America. Obviously, a large number of government officials and corporate leaders are sorely troubled by what they consider to be illegal disclosure activities; they want those involved in the leaks prosecuted and punished. On the other hand, Julian Assange, the CEO of WikiLeaks, considers its activities as key to maintaining democratic systems that rely upon an informed citizenry. Assange considers himself a valiant defender of the rights of citizens to demand a transparent society. He is convinced that the charges leveled against him for sexual harassment are backed by powerful organizations that are afraid of the consequences of full disclosure of their activities. The battle lines are drawn in this conflict, and it is uncertain how it will be resolved. It seems inevitable that similar cases of disclosure of sensitive data will occur in future. What is clearly needed is a new definition of the balance between privacy and transparency, for ICTs have roiled the waters to the extent that the old balance is no longer functional.

CULTURE WARS

As with most of the topics discussed in this book, culture wars have a long history. The concept of *culture wars* has been applied in many situations and for many purposes. The most general definition of the concept is the disagreement that arises among competing groups that espouse contrasting economic and political agendas. The conflict arises from those who feel strongly that only their ideology should prevail. Given this most general conception, it is reasonable to assume that such conflicts have been a historical constant in all societies from the hunting and gathering tribes to the agricultural communities, and continuing up to the contemporary megalopolis. Indeed, Adam Smith's analysis of the division of labor included the concept that a person whose labor was more valued would reap greater benefits than those whose labor was less valued. The miller who ground the wheat would be more affluent than the farmer who grew the crop, because the creator and operator of the mill had used special skills and expertise. Thus were the beginnings of social class stratification and the foundation of all subsequent culture wars.

Over the many centuries before and since the Industrial Revolution, the culture wars have taken on a variety of substantive dimensions. However, at the common core of the culture wars are the different perceptions of those who exercise power over others and those who wish to change the balance of power to promote the values that they desire for the future structure and functioning of society. Today, the issues that define the culture wars in North America and Europe include deficit spending, creationism, gun control, stem-cell research, sex education, gay rights, free markets, abortion, and immigration policies. Discussion, disagreement, and debate on such issues are certainly not new, and many of the proponents of different positions roughly exhibit the characteristics of conservatives or liberals or Republicans or Democrats. However, many have observed that the discourse from both sides of these issues has become markedly more strident in recent years. Indeed, on some issues such as abortion, opponents have turned to violence resulting in deaths to emphasize the zealotry of their beliefs. In the U.S., the recent deadlocks of legislative bodies at both the national and local levels are a consequence of the extreme positions of elected officials in the culture wars.

ICTs are contributing to the polarization that now characterizes the cultural wars. The plethora of blogs, websites, listservs, and discussion groups devoted exclusively to one side or the other in the debates increases the vehemence and negativism expressed. For example, if one wants to find materials promoting white supremacy, a multitude of websites support that position. Never before was it possible for so many people readily to find such extremist content. Both the volume and frequency are escalated by ICTs. Furthermore, the anonymity of content posted on the Internet encourages many to promote even more extreme positions. Extremists are not new in political arenas. Indeed, pamphleteers espousing all sorts of causes are almost a tradition. Recall the virulent publications that preceded and accompanied all the major conflicts in U.S. history, from the Revolution to the Civil War, both World Wars, Vietnam, and the current conflicts in the Middle East. Clearly, radio and television talk show hosts are major contributors to the divisiveness of the culture wars. Most of those broadcasts are further distributed as downloaded files over the Internet. Hence, the harshness of the culture war debates is greatly enhanced by ICTs.

GLOBALIZATION

Globalization is also a term that has accumulated many definitions, economic analyses, historical accounts, and controversies. And each of these perspectives could produce multiple volumes. This discussion employs the definition proposed by the International Monetary Fund consisting of four components of internalization: trade, investments, migration, and knowledge transfer. It goes without saying that the implementation of the globalization processes would be impossible without ICTs. Regardless of whether globalization focuses on world economic trade, facilitating economic growth in developing nations, expanding educational opportunities, protecting the environment, population growth, or the establishment of worldwide peace, the vehicle that enables globalization is enhanced communication among all peoples of the world. The paramount policy question for the future is how will globalization and the ongoing transition to the informational state proceed and how can it be facilitated?

Thomas Friedman, author and *New York Times* columnist, has formulated an expanded and provocative conception of globalization. Through his recent books and op-ed columns, Friedman has explored and explained a historical framework of globalization. He claims that international economic and political relationships have progressed through three stages, which he labels as *countries, nations,* and *individuals.* These generalizations oversimplify many historical and economic events, but they do provide a useful framework for understanding broad trends and the dynamics of globalization. Prior to World War II, international relations were based on the interactions of countries, Friedman's stage one. The conflicts that occurred involved one or several groups of countries that fought over the location of borders or access to trade routes or ownership of the natural resources or labor supplies of colonized areas. World War II marked the beginning of Friedman's stage two, and it pitted a group of nations known as the Allies against another group known as the Axis. The outcome halted the expansionist aspirations of both Germany and Japan. The post-war era ushered in the Cold War period when most nations in the world aligned themselves with either the U.S. or the Union of Soviet Socialist Republics led by Russia. The essence of the nation stage was expressed in the Berlin Wall, which separated the two super powers and their allies.

The Internet has ushered in the third stage, in which individuals operate in both economic and political arenas independent of governments. The Internet enabled commerce and communication among people and corporations that was previously impossible. Friedman identifies the Web as the essence of the third and current stage. In this view, globalization and ICTs are both the cause and effect of the problems and potentials of the world today. For example, Al Qaeda and Osama Bin Laden were able, independently of any government, to use the Internet to organize and execute multiple acts of terrorism with dire consequences. Al Qaeda was purportedly behind the disastrous attacks in several locations in the U.S., London, and Barcelona, among others. The U.S. government launched an unsuccessful missile attack on Bin Laden in Afghanistan in 1998. Yet Bin Laden and Al Qaeda continued to operate worldwide for another 13 years before he was assassinated. This example illustrates how ICTs can enable activities by extremist groups that have caused significant death and destruction and, perhaps more importantly, widespread fear. It is interesting to speculate if further expansions of the capabilities of ICTs will enable a transition to the informational state. And, if so, what will be the configuration of the social, political, and economic changes that will evolve?

A FINAL THOUGHT

Promises have repeatedly been made that this book would not include predictions about the future of computers, networks, and applications. However, this concluding note will now violate the promise and refer to some possible directions of new developments. The intention here is to provide a context in which readers might reflect on the future directions of social change that could be enabled with new and more powerful computation devices. An interesting, recently published book, *Natural Computing: DNA, Quantum Bits, and the Future of Smart Machines*, by Dennis Shasha, professor of computer science at New York University, and Cathy Lazere, a freelance journalist, explores such possibilities as using biological materials in computers, designing machines that can analyze and correct themselves, and creating analog devices that measure changes in physical characteristics. Such innovations could create computational capacities that would be smaller, faster, and more reliable by orders of magnitude. The economic, political, and sociological impacts of such capabilities would be enormous. It is tempting to wish to live for another hundred years to witness how ICTs will influence the next stages of societal evolution.

This book provides a framework exploring some of the major trends of human societal evolution. ICTs influence those trends, escalating some and redirecting others. Undeniably, ICTs are instruments for enhancing both global welfare and harm. It is up to us to decide and provide direction.

Appendix A
Design and Architecture of Digital Computers

This appendix briefly discusses the basic design and logic of digital computers. In this appendix and throughout the book, the focus on information communication technology (ICT) systems is limited to digital computers and networks and excludes analog devices. Analog computers are a different type of machine in which a continuously varying physical property, such as velocity, temperature, voltage, or pressure, is used for the purpose of calculation. Current use of analog computers is mostly limited to fairly narrow areas of scientific or engineering research, for example, determining the optimum shape of the wings of airplanes or the hulls of ships. The vast majority of computers and networks in use today are digital. However, some areas of experimental research involving the use of biological and quantum components of computational devices may, in the not-too-distant future, blur the distinction between digital and analog machines. Nevertheless, the analyses in this book exclusively address digital systems.

The goal of this appendix is to establish a basic understanding of the elementary components and capabilities of computers. All too frequently, computers are portrayed by the mass media, and unfortunately by some ICT professionals, as black boxes that mysteriously work wonders which can be understood only by the most gifted individuals with extensive and esoteric training. Not only does this depiction inspire unwarranted awe, but it also confuses people as to what might reasonably be expected from computer applications in the near future. This section dispels some of that unnecessary confusion and allows the reader to distinguish between what computers might readily accomplish and those applications that must await further fundamental development in other branches of science and technology before an ICT system will be feasible.

The initial step in understanding digital computers requires an examination of the ways in which data are represented internally. The most elementary component of a digital computer is a micro-miniature device that, at any given time, exists in only one of two possible states. This component is called a bit, or a binary bit, referring to its two-state character. The actual material used to construct these elementary components has changed over the years from vacuum tubes to ferroelectric rings to silicon chips. Scientists are always looking for other possible materials to use for bits in computers. The objective of their efforts is to find materials that will decrease cost, increase miniaturization, permit greater speed, require less power, generate less heat, and enhance performance and overall reliability. Except for recognizing that these enhanced performances are a consequence of using new materials for binary storage, their substance is not central to this discussion. Rather, the focus here is on conceptual design and logic.

The two binary states of the basic bit correspond to the clockwise or counterclockwise flow of the electromagnetic force in the chip. The direction of that flow can be detected and changed while the computer is operating. The flow is referred to as *polarity*, and it represents two states, zero and one. This simple binary capability allows the representation of any number system as well as all the letters and punctuation marks used in the many languages throughout the world. Furthermore, the binary bit system can be used to represent all the basic arithmetic and logical operations. As a common practice, contiguous bits are usually collected into groupings. Eight bits are referred to as a byte; bytes are combined into words; and 16-, 32-, 64-, and 128-bit words are common. It is important to keep in mind that the simple binary bit representing either a zero or a one is the basic unit of all machines.

At any given time, the memory of a computer is filled with strings of binary zeroes and ones. Some of these strings are bytes and words representing data, either numerical or textual. Other strings contain binary codes for instructions which the computer can execute, such as add, subtract, or compare. The instructions also include the addresses or locations within the computer memory of the *operands*, or the objects of execution of the instruction, such as the numbers to be added, subtracted, or compared. Instructions are the steps in a program to accomplish a specific task. These steps are called the *algorithm*. The processes of designing the algorithm and writing the instructions make up *computer programming*.

Figure A.1 depicts the basic architecture of a digital computer. The figure represents all the major components and input/output devices of any digital computer, including notebooks, laptops, desktops, work stations, minicomputers, mainframes, or even super computers. Regardless of size, speed, or complexity, all digital computers share this same fundamental design.

The central processing unit (CPU), in the middle of the figure, is literally the heart of the machine. The CPU contains the circuitry that interprets and executes instructions. It is divided into two sections. The first is the arithmetic and logic unit (A & L in the figure), which executes the instructions by interpreting the binary code, retrieving data from memory, and

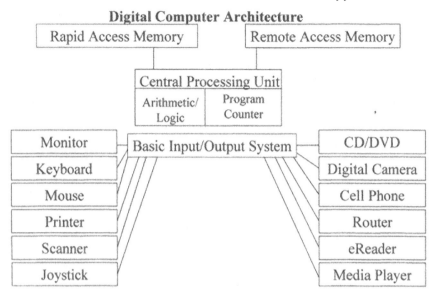

Figure A.1 Digital computer architecture.

then storing the results. The other section of the CPU is the program counter unit, and its primary function is to keep track of the next instruction to be executed.

The CPU interacts with the two other major components of the computer, memory and the input and output devices. As the figure depicts, rapid and remote access are the two types of memory. The former refers to the words in memory that are immediately accessible to the arithmetic and logic unit as instructions and operands. The remote access memory, as the name implies, is stored on a variety of devices that may include an internal hard disk or a number of different external devices such as another hard drive; a memory stick; or even a remote storage unit in a distant location, increasingly referred to as a *cloud*. Through the basic input and output system (BIOS), which is itself a complicated program, the computer can receive and send binary bit patterns to a wide variety of devices. Figure A.1 lists most of the current major input/output (I/O) devices. However, these devices are the most rapidly changing components of computers.

It would be cumbersome, and, in many cases, impossible, to write the instructions of all computer programs as strings of binary code. For this reason, many higher-level languages have been created to make the task of writing programs more manageable. The plethora of computer higher-level languages has often been compared to the ancient Tower of Babel.

At the same time that all the software advances in operating systems and higher-level languages were occurring, important breakthroughs were happening in computer hardware. Many histories of these developments are available, and only the main trends of increasing power and reliability and decreasing size and cost are mentioned briefly here. CPUs with the same architecture described previously were placed in the earliest large mainframe machines and later in smaller minicomputers. By the early 1980s, personal computers (PCs) appeared and drastically changed patterns of use. This machine made available in a box that fit on a desktop all the power that just 20 years earlier was available in a mainframe computer that occupied several thousand square feet of space. The first PCs sold for around $5,000, approximately 1 percent of the price of the old mainframes. A much more powerful PC now sells for several hundred dollars, and that is approximately one-tenth of 1 percent of the cost of a mainframe computer in the 1960s.

As the price of PCs declined, the numbers of users escalated. Most of these people were new to computers, and they found the operating systems and programming languages awkward and cumbersome. To address this problem of usability, the graphical user interface (GUI) was developed at the research division of the Xerox Corporation in Palo Alto, California. The GUI substituted a pointing device, a mouse, and icons on pull-down menus for the lines of the text commands of the PC programming languages. The GUI was a major design advance, for it allowed the use of the PC without entering program instructions on the keyboard. It also introduced the icon as a tool for executing many machine language instructions. For example, moving the mouse over a small picture of a printer and clicking activates all the instructions necessary to print a document. Or clicking on an icon of a trashcan executes all the instructions to delete a file from a storage device. The use of icons for interacting with the computer may open a new mode of communication.

This very brief overview of design and logic is intended to provide the basis for appreciating the simplicity of the basic components of digital computers. When these simple components operate at speeds of millions of instructions per second, it becomes possible to use these capabilities for enormously complicated arithmetic and logical calculations.

Appendix B
Computer Networks

As the speed and power of computers increased, time-sharing systems were developed to maximize their effective use. First perfected at Dartmouth College and the Massachusetts Institute of Technology (MIT), early time-sharing systems used teletype terminals to connect many users to the central processing unit (CPU) of a computer with large local and remote storage capacities. The user's experience was similar to that of having the entire computer's power immediately and exclusively available. Time-sharing systems spread quickly, and personal computers (PCs) soon replaced the Teletype terminals. These time-sharing systems provided the foundation for building the networks that are so ubiquitous today.

The first time-sharing systems were actually local area networks (LANs), for example, in or near Dartmouth College in Hanover, New Hampshire, or Harvard University or MIT in Cambridge, Massachusetts. The first wide area network (WAN) was developed by the U.S. Department of Defense (DOD) to allow geographically distributed military installations to communicate with one another and exchange data files. The DOD's Advanced Research Projects Agency (DARPA) supported the development of the network, and the net became known as ARPANET. Universities and research institutes whose activities were supported by DOD also joined the net. Eventually, the traffic on the net became so heavy that the military units withdrew from ARPANET and established their own MILNET. ARPANET then evolved into the current Internet by merging with other networks such as BITNET, Usenet, and NSFnet.

Figure B.1 depicts the basic architecture of computer network services. The CPU of a digital computer is connected to the Internet through a variety of different data transmission systems. Devices called *modems* translated the binary bit patterns to the network. They converted the digital patterns into analog signals for transmission over the copper wires then used in most commercial telephone systems. However, the transmission rates were painfully slow, and high-speed or broadband systems were developed to make access faster and more reliable. Cable, fiber optic service, and increasingly wireless transmissions systems replaced modems, and their faster speeds made Internet access more manageable and appealing to many people. The

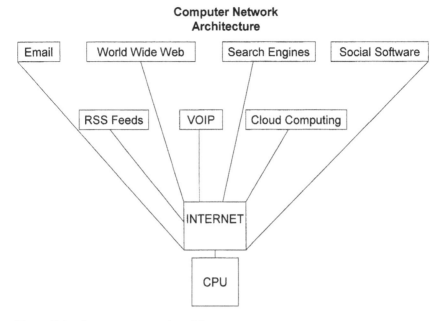

Figure B.1   Computer network architecture.

specific differences among modems, cable, FIOS, and Wi-Fi are not relevant to this discussion. The important point is that transmission speed increased dramatically, and the Internet was transformed from a limited, text-based information exchange system to a rich variety of audio and video capabilities and services that rapidly appealed to users around the world.

The Internet continues to grow into a network consisting of millions of commercial, government, and educational computer sites and networks. These networks exchange files by using packet switching and common data transfer protocols. The networks that comprise the Internet reside on a wide variety of computers ranging from handheld devices to supercomputers. The global-spanning networks are interconnected with a variety of devices, including copper wires, broadband cables, wireless transmitters and receivers, and satellite communications systems. No central, international agency oversees the Internet. Therefore, no one knows for certain how many Internet connections exist. More than a billion people are estimated to access the Internet every day. China alone has more than 400 million Internet users. Before briefly describing the many and varied activities enabled by this global-spanning Internet, it is worth recalling from the earlier description of the design and logic of digital computers that all these vast Internet exchanges consist of strings of binary bits representing only two symbols, zero and one.

Figure B.1 illustrates seven of the most commonly used capacities of the Internet: electronic mail or email, the World Wide Web (WWW), search engines, social software, Really Simple Syndication (RSS) feeds, Voice over Internet Protocol (VoIP), and cloud computing. Each of these Internet activities is described briefly to provide an overview of the variety of information exchanges enabled by this group of information communication technologies (ICTs). Detailed discussions of the origins, growth, and distribution of the first four of these Internet activities are included in the analyses in previous chapters of this book. The remaining three are briefly described next.

Really Simple Syndication (RSS) feeds are programs subscribed to by users who receive notices about new or updated Web pages on a specified topic. The RSS aggregates updates to Web pages and informs subscribers of the new materials. Although there are multiple and conflicting formats for RSS systems, the practice of news aggregation is widespread and growing rapidly. Many users find it efficient to set up RSS feeds to keep them informed about new developments in one or a number of fields. These systems are analogous to having a full-time reader scan a large number of current publications, including magazines, journals, newspapers, and even miscellaneous ephemera and bring to your attention all materials of potential interest. If nothing else, RSS feeds certainly enhance the efficiency and scope of subscribers' reading behaviors.

Voice over Internet Protocol (VoIP) is enabled by the Internet and is experiencing a rapid expansion. VoIP uses the Internet to allow users to make telephone calls at low or no cost. Vonage is a growing commercial VoIP service. With a special modem and a monthly fee, Vonage subscribers can contact others around the globe. If the person called is on a land telephone line, there is no charge. If the person is on a mobile or cell phone, there is a minimal charge. Skype is another rapidly expanding commercial VoIP service. It enables visually enhanced conference calls. The participating parties must have a video camera attached to their computers. Visual telephones are an innovation that many companies have promised for years, but the Internet has finally made them a viable commercial reality. If the number of users of VoIP systems continues to grow, the major suppliers of telephone services, including landlines, cable, and microwave, may find their customer bases starting to diminish. These companies will likely offer competing services.

Cloud computing is the ICT-enabled innovation that may have the most far-reaching impact in the future. In this implementation, hardware, software, and data are all located at a remote computing facility that is accessed via the high-speed broadband connections now available through the Internet. Referring back to Figure A.1, all the components of the CPU and memory reside on a remote computer system that can be located anywhere on the globe or on a satellite circling the globe. The input/output devices in Figure A.1 are located in front of the user, who directs the operations of the software on the remotely stored data. All this interaction among the user, the hardware, the software, and the databases is mediated through the Internet. Cloud computing can be economically efficient, because the user pays only for those capacities required for a particular task. Cloud computing is analogous to utilities that provide electrical power or water to their subscribers. Such utilities are efficient and eliminate the need for individuals or families to maintain their own electrical power generators or provide their own water wells. This comparison raises an interesting question concerning regulation. Government regulators carefully oversee the operations of most public utilities to ensure quality control and fiscal accountability. As cloud computing continues to expand, will it invite a similar regulatory oversight?

The Internet has greatly expanded the manifold ways that digital computers can be utilized to accomplish many tasks that 50 years ago were completed solely with human labor. For example, all the operations associated with banking were once accomplished by clerks who copied numbers and tallies into ledgers and accounting books. Internet-supported banking now permits electronic fund transfers for deposits, payments, reconciliations, and even loan applications. All these functions are executed with efficiency and accuracy with little, if any, human input or oversight. Furthermore, the Internet has created new means for accomplishing tasks that previously would have been considered too onerous and expensive even to attempt. For example, 50 years ago, no one would have considered hiring enough readers of all the documents published in the previous year on the topic of immigration policies. It would have been an impossible project. Today, this would be a trivial job employing RSS feeds. The reader is again cautioned to keep in mind that all these functions are accomplished with extensive and rapid manipulation of strings of binary bits representing nothing more than zeroes and ones.

Until recently, most Internet use has been with computers connected to servers by wires, most commonly the same wires used for telephone networks. However, wireless networks are now expanding rapidly and facilitating greater use of mobile devices. More manufacturers are now producing smart phones that incorporate many of the features of personal computers. These handheld devices and mobile computing are usually less expensive to purchase and frequently somewhat easier to use than desk or laptop personal computers.

These wireless networks that facilitate mobile computing are beginning to address the inequities of the digital divide. In many developing countries, more cell phones than personal computers are in use. Of course, wireless networks are not evenly distributed around the world. They tend to be found in densely populated urban areas, such as Hong Kong, Helsinki, and Seoul. Many urban and rural areas plan to create wireless networks. Clearly, mobile computing will continue to proliferate, and now pundits refer to such applications as m-learning, m-health, and m-government. Mobile commuting may well be the next big thing.

Biographical Notes

CHAPTER 1

In this introductory and subsequent chapters, the late Daniel Bell's work provides the intellectual context. His book, *The Coming of the Post-Industrial Society: A Venture in Social Forecasting*, was first published in 1973 by Basic Books. He subsequently wrote two Forewords, one in the 1976 edition and the second for the 1999 edition. Readers are encouraged to read the latest edition to assess the breadth and depth of Bell's social forecasting acumen. ISBN: 0465012817.

The section describing the scope of this investigation draws heavily on the definitions of the words *information*, *communication*, and *technology* as defined in the venerable *Oxford English Dictionary*. This invaluable resource is now available online, but access is limited to individual or institutional subscriptions. However, most academic and large public libraries offer access to their patrons.

Details describing the iSkills test produced and distributed by Educational Testing Service are available online at the following URL (2013): ets. org/iskills/about

CHAPTER 2

This chapter cites the work of many of the best known social scientists who have written extensively on various dimensions of theory of the structure and functioning of society. Descriptions and excerpts of their work appear throughout the sociological literature. An excellent source for additional information on the writings of all the people mentioned in Chapter 2 is the two-volume set *Theories of Society*, edited by Talcott Parsons, Edward Shils, Kaspar D. Naegele, and Jesse R. Pitts, published by The Free Press of Glencoe. Although published in 1961, this set is still one of the best sources for materials by the founders of sociological theory.

The citation for Daniel Bell's *The Coming of Post-Industrial Society* is included in the bibliographic notes following Chapter 1.

Two contemporary authors are also mentioned in this chapter. They are: James J. O'Donnell, *Avatars of the Word: From Papyrus to Cyberspace*, Harvard University Press, 1998; and Albert Borgmann, *Holding Onto Reality: The Nature of Information at the Turn of the Millennium*, University of Chicago Press, 1999.

Readers are encouraged to look at Naomi Baron's highly readable book *Always On: Language in an Online and Mobile World*, Oxford University Press, 2008.

Braman's prescient work is detailed in *Change of State: Information, Policy, and Power*, The MIT Press, 2006.

CHAPTER 3

Since the original publication in 1651, *Leviathan, or The Matter, Form & Power of a Common-Wealth Ecclesiastical and Civil* by Thomas Hobbes has been edited and reprinted numerous times. A recent version with an introduction by the contemporary philosopher J. C. A. Gaskin was published in 1996 by Oxford World Classics.

Two books that extend the brief discussions of normative order presented in this chapter are Dennis H. Wrong, *The Problem of Order: What Unites and Divides Society*, The Free Press, 1994; and William F. Ogburn, *Social Change: With Respect to Cultural and Original Nature*, Delta Books, 1966. The latter introduces the concept of *cultural lag*, which is being resurrected by many instances related to ICTs.

For readers interested in more comprehensive analyses of normative order and the role of ICTs, two authors are recommended. Manuel Castells, professor of sociology at University of California, Berkeley, has completed a three-volume analysis entitled *The Information Age: Economy, Society, and Culture, Volume One: The Rise of the Network Society; Volume Two: The Power of Identity; and Volume Three: End of Millennium*, Blackwell Publishers, 1996. A more concise analysis is contained in his *The Internet Galaxy: Reflections on the Internet, Business, and Society*, Oxford University Press, 2001. The second author is Yale Law School Professor Jochai Benkler, and his book is *The Wealth of Networks: How Social Production Transforms Markets and Freedom*, Yale University Press, 2006.

The theoretical literature on social change and deviance is discussed extensively in the materials referenced in Chapter 2. An empirical investigation of deviance in the early New England colonies is Kai T. Erikson, *Wayward Puritans: A Study in the Sociology of Deviance*, John Wiley & Sons, 1966.

The classic introduction to the concept of the division of labor can be found in Adam Smith's, *An Inquiry Into the Nature and Causes of the Wealth of Nations*, Methuen and Co., Ltd., edited by Edwin Cannan, 1904, fifth edition. Smith, an eighteenth-century philosopher and economist, is widely known today as the first advocate of free markets. However, his contributions to understanding the division of labor influenced many of the early sociologists. A recent study of how ICTs are influencing the contemporary division of labor and attendant markets is reported in Frank Levy and Richard J. Murnane's *The New Division of Labor: How Computers Are Creating the Next Job Market*, Russell Sage Foundation and Princeton University Press, 2004.

For a review of the problems and potentials of electronic voting see the reports issued by Professor Edwin W. Felten and his students at the Princeton University Center for Information Technology Policy at citp.princeton.edu

CHAPTER 4

The existing literature on the issues of ownership and privacy is vast and continues to expand vigorously. However, the person who has probably contributed most to deliberations on these questions and open access is Professor Lawrence Lessig of Harvard Law School. His first book, *Code and Other Laws of Cyberspace*, Basic Books, 1999, raised concerns that the anticipated freedom of access which early on accompanied the creation of the Internet was in danger of being curtailed. The commercial interests were supported by government revisions of copyright laws to obtain and restrict access to intellectual and creative products. They wish to profit by charging fees to access the content of cyberspace. His second book, *The Future of Ideas: The Fate of the Commons in a Connected World*, Random House, 2001, continues his analysis of the eroding of open access to knowledge promoted by corporate interests and enabled by government regulation of copyright and patents. In his third book, *Free Culture: How Big Media Uses Technology and the Law to Lock Down Culture and Control Creativity*, The Penguin Press, 2004, makes a convincing case for limiting the power of companies that prevent access to materials and ideas that previously were in the public domain. Readers are urged to become familiar with Lessig's works, for they will be an important part of a coming public debate over questions of ownership, privacy, and open access.

The summary of the research on privacy in the trucking industry draws from Karen E. C. Levy, "The Automation of Compliance: Techno-Legal Regulation in the U.S. Trucking Industry," forthcoming doctoral dissertation, Princeton University, 2013.

An early and still relevant collection of papers on privacy that were delivered at a 1970 conference is contained in *The Information Utility and Social Choice*, edited by Harold Sackman and Norman H. Nie, American Federation of Information Processing Societies Press, 1970.

Two interesting volumes that shed additional light on the use of ICTs in religious settings are Robert Glenn Howard, *Digital Jesus: The Making of a New Christian Fundamentalist Community on the Internet*, NYU Press,

2011; and Elizabeth Drescher, *Tweet If You Love Jesus: Practicing Church in the Digital Reformation*, Morehouse Publishing, 2011.

A recently published collection of original essays on criminal behaviors enabled by ICTs is *Cyber Criminology: Exploring Internet Crimes and Criminal Behaviors*, edited by K. Jaishankar, CRC Press, 2011. Jaishankar is a leader in the international community of scholars who are investigating these new criminal patterns. He is a professor of criminology at Manon Sundaranar University in India. The 21 essays written by scholars from all over the world cover such topics as gambling, sex offenses, terrorism, and piracy.

CHAPTER 5

The Pew Internet & American Life Project is part of the Pew Research Center's activities. The project collects data through nationwide telephone surveys on a wide variety of topics having to do with ICTs and education, families, communities, and health. Reports from the project are available at (2013) www.pewinternet.org/About-Us.aspx

See Harriet B. Presser, *Working in a 24/7 Economy: Challenges for American Families*, Russell Sage Foundation, 2003.

Marc Prensky's work on natives and immigrants and his promotion of computer games as learning devices has generated considerable debate and discussion. Readers are urged to review his work at (2013) www.marcprensky.com

Nicholas Carr's article "Is Google Making us Stupid?" appeared in the July/August 2008 issue of *The Atlantic*. More recently, he extended his argument in *The Shallows: What the Internet is Doing to Our Brains*, Norton, 2011.

Mahoney's King's College lecture is reprinted in Michael Sean Mahoney, *The Histories of Computing*, edited and with an introduction by Thomas Haigh, Harvard University Press, 2011.

A brief history of the PLATO project is available online at (2013) www.thinkofit.com/plato/dwplato.htm

Up-to-date information on systems thinking applications in schools is available online at (2013) http://www.clexchange.org/

The Babson-Sloan surveys are available online at (2013) http://sloanconsortium.org/publications/survey/class_differences

Readers who wish to learn more about intelligent tutoring systems would do well to read *Intelligent Tutoring Systems: An Historic Review in the Context of the Development of Artificial Intelligence and Educational Psychology*, by Mark Urban-Lurain, is available at (2013)http://www.cse.msu.edu/rgroups/cse101/its.htm

The report of the Commission on the Humanities and Social Sciences of the American Academy of Arts & Sciences is available at (2013) www.humanitiescommission.org

The collaborative learning project is reported in Douglas Thomas and John Seely Brown, *A New Culture of Learning: Cultivating the Imagination for a World of Constant Change*, self-published, 2011.

Although there is now ample literature on interactive teaching, Mazur's book is still a foundation work. Eric Mazur, *Peer Instruction: A Users' Manual*, Prentice Hall, 1997.

The Father Guido Sarducci parody on higher education is available at (2013) http://www.youtube.com/watch?v=kO8x8eoU3L4

The Christensen and Horn article is "Colleges in Crisis: Disruptive Change Comes to American Higher Education," *Harvard Magazine*, July–August 2011, Volume 113, Number 6, pp. 40–43. The article applies the disruptive innovation perspective developed by Christensen and Horn in schools in the book, *Disrupting Class, Expanded Edition: How Disruptive Innovation Will Change the Way the World Learns*, Clayton Christensen, Curtis W. Johnson, and Michael B. Horn, McGraw Hill, 2011.

CHAPTER 6

The websites for the three large-scale database projects are (2013): http://www.norc.uchicago.edu/GSS+Website/, http://nces.ed.gov/surveys/nels88/, and http://memory.loc.gov/ammem/index.html

Visit the Culturomics Project at its home page at (2013) http://www.culturomics.org

The original Vannever Bush article can be found at (2013) http://www.theatlantic.com/magazine/archive/1945/07/as-we-may-think/3881/

The home page website for the various databases under the auspices of the Wiki Foundation is (2013) http://www.wikipedia.org/

The original article by Jeff Howe on crowd sourcing is online at (2013) http://www.wired.com/wired/archive/14.06/crowds.html

The fascinating history of the production of the first edition of *The Oxford English Dictionary* is recounted in Simon Winchester's *The Professor and the Madman: A Tale of Murder, Insanity, and the Making of the Oxford English Dictionary*, HarperCollins, 1998.

The materials discussing the histories of libraries are drawn from Lionel Casson, *Libraries of the Ancient World*, Yale University Press, 2001.

The Association of Research Libraries regularly produces reports discussing issues confronting libraries. These publications can be accessed online at (2013) http://www.ala.org/ala/mgrps/divs/acrl/publications/whitepapers/whitepapersreports.cfm

The volume that includes the most comprehensive discussion of the changing nature and role of information in society is the recently published book by James Glieck, *The Information: A History, a Theory, a Flood*, Pantheon Books, 2011.

CHAPTER 7

The U.S. data on retail sales, online purchases, and advertising expenditures all come from the economic section of the Census Bureau at (2013) http://www.census.gov/econ/index.html

The data on Amazon revenues and employees are taken from the 2010 form submitted to the U.S. Securities and Exchange Commission.

The data on e-Bay revenues and employees are taken from the company's report to shareholders in January 2011.

The original article by Chris Anderson was published in *Wired* magazine in 2004 and is available online at (2013) http://www.census.gov/econ/index.html. He expanded the concept into a book, *The Long Tail: Why the Future of Business Is Selling Less of More*, Hyperion, 2006.

The Hardin article can be accessed in the *Science* magazine archives with this information: "The Tragedy of the Commons," Volume 162, Number 3859, pp. 1243–1248, 1968.

The Meadows book is by Donella H. Meadows, Dennis L. Meadows, Jorgen Randers, and William W. Behrens, *The Limits to Growth: A Report of the Club of Rome's Project on the Predicament of Mankind*, Universe Books, 1972.

The interview with Summers and Immelt is summarized at (2013) http://seekingalpha.com/article/148477-larry-summers-and-jeff-immelt-preparing-for-a-post-consumer-economy

CHAPTER 8

For a concise history of ancient postal services, visit the website at (2013) http://www.lookd.com/postal/history.html, and for the U.S., see the pdf file of an 84-page illustrated history of the USPS at (2013) http://about.usps.com/publications/pub100.pdf

A history of the development and deployment of the telegraph is contained in Tom Standage's *The Victorian Internet: The Remarkable Story of the Telegraph and the Nineteenth Century's On-Line Pioneers*, Walker, 1998. Claude S. Fischer's social history of the telephone is *America Calling: A Social History of Telephone to 1940*, University of California Press, 1992.

Dava Sobell chronicles the fascinating story of the accurate measurement of time on seafaring vessels in *Longitude: The Story of a Lone Genius Who Solved the Greatest Scientific Problem of His Time*, Penguin Books, 1995.

Clark Blaise in his book *Time Lord: Sir Sandford Fleming and the Creation of Standard Time*, Pantheon Books, 2000, recounts the remarkable story of the man who persuaded the nations of the world to adopt a standard for measuring time. A definitive account of the measurement of time since the Industrial Revolution is contained in the book by the late David S.

Landes, *Revolution in Time: Clocks and the Making of the Modern World*, Belknap Press, 1983.

An early and lucid description and analysis of virtual communities was reported by Howard Rheingold in *The Virtual Community: Homesteading on the Electronic Frontier*, Addison-Wesley, 1993.

For a fascinating account of his experiences in various virtual communities, see Julian Dibbell's *Play Money: Or, How I Quit My Day Job and Made Millions Trading Virtual Loot*, Basic Books, 2006.

Tom Boellstorff, an anthropologist and professor at the University of California, Irvine, writes a most interesting and informative book, *Coming of Age in Second Life: An Anthropologist Explores the Virtually Human*, Princeton University Press, 2008. He effectively employs his anthropological skills in fieldwork in Second Life and provides a most informative description of life in a virtual community.

More information about social science research on virtual communities can be obtained from the activities and publications of the University of Toronto-based Netlab, under the direction of Barry Wellman, professor of sociology. Visit the website at (2013) http://homes.chass.uNtoronto.ca/~wellman/

The website of the Association of Internet Researchers is (2013) http://www.aoir.org/

CHAPTER 9

Readers are urged to view a YouTube video demonstrating Dr. Rosling's historical analysis of international public health over the past century. The video can be viewed at (2013) http://www.youtube.com/watch?v=jbkSRLYSojo

The work of Professor Shneiderman and his colleagues is available for summary review online at (2013) http://www.cs.umd.edu/~ben/

For a recently published collection of original essays on spatial humanities, see *The Spatial Humanities: Geographical Information Systems (GIS) and the Future of Humanities Scholarship*, edited by David J. Bodenhamer, John Corrigan, and Trevor M. Harris, Indiana University Press, 2010.

The three Thomas Friedman books most relevant to this discussion are *The Lexus and the Olive Tree: Understanding Globalization*, Farrar, Straus, & Giroux, 1999; *The World Is Flat: A Brief History of the Twenty-First Century*, Farrar, Straus, & Giroux, 2005; and *Hot, Flat, and Crowded: Why We Need a Green Revolution—And How It Can Renew America*, Farrar, Straus, & Giroux, 2008.

The future directions of ICT hardware and software are described in Dennis Shasha and Cathy Lazere, *Natural Computing: DNA, Quantum Bits, and the Future of Smart Machines*, W. W. Norton & Company, 2010.

APPENDIX A

For a more detailed discussion of the history and architecture of ICTs readers are encouraged to see Martin Campbell-Kelly and William Aspray, *Computer: A History of the Information Machine*, second edition, Westview Press, 2004.

Index

Printed in the United States
by Baker & Taylor Publisher Services